An Herbal Selection

ALFALFA

Medicago sativa L.
(family Fabaceae)

Description: A nutritious fodder plant with violet flowers. Its roots extend some three and one–half inches below the soil.

Range: First cultivated in Persia, it was taken to Greece in the fifth century B.C. and to Spain in the eighth century A.D. Spaniards introduced it to North and South America.

Application: No human curative properties, but excellent as feed for all forms of livestock.

Comments: Large doses of alfalfa seeds or sprouts ingested by humans can result in irreversible blood abnormalities, and monkeys have been found to have lupus (connective tissue disease) after being fed alfalfa seeds and sprouts.

ANGELICA

Angelica archangelica L.,
Angelica atropurpurea L.
(family Apiaceae)

Description: A tall, perennial herbaceous plant containing a very pleasant–smelling volatile oil.

Range: Europe and North America.

Application: Besides its use as a flavoring in confections and liqueurs, it has been recommended as an antiflatulant, a diuretic, a diaphoretic (sweat producer), and a counterirritant. There is no proof that it is particularly effective in any of these applications.

Comments: Studies have shown that various compounds contained in the volatile oil are acutely toxic and mutagenic in laboratory animals.

ARNICA

Arnica spp.
(family Asteraceae)

Description: A perennial herb with orange–yellow daisy–like flower heads.

Range: Native to the meadows and mountainous regions of Europe and western North America.

Application: External application has been recommended to reduce the inflammation and pain of bruises, aches, and sprains. Studies on small animals have shown that internal application has a toxic effect on the heart and causes very high blood pressure.

Comments: There is considerable rationale for the external application of arnica to reduce inflammation and pain, although some sensitive persons who use it this way may experience contact dermatitis. Internal use of the drug has been shown to be toxic and must be discouraged.

BORAGE

Borago officinalis L.
(family Boraginaceae)

Description: An annual herb with rough hairy stems and leaves. The inflorescence (flower cluster), a coiled spiral that unrolls and straightens from the base as the flowers open, is the hallmark of the family.

Range: More than 2,500 species are found in the temperate and subtropical regions of the world.

Application: Used as a culinary herb since the Middle Ages. Folk remediesclaimed it could instill courage and dispel sadness. In truth, it has practically no positive effect on the body, barring mild expectorant, diuretic, and constipatory effects.

Comments: Chronic consumption of the plant or seed oil should be carried out only under medical supervision, unless the product has been certified to be free of all unsaturated pyrrolizidine alkaloids (UPAs).

BURDOCK

Arctium lappa L. and *Arctium minus* (Hill) Bernh.
(family Asteraceae)

Description: Coarse, rough biennial weeds, characterized by bristly burrs, or bracts, which adhere to clothing and animal fur.

Range: Native to Europe; naturalized in the United States.

Application: Mainly in folk medicine as a blood purifier; also thought to aid in treating chronic skin maladies. Some diuretic and diaphoretic properties.

Comments: No conclusive clinical tests have been done to verify the effectiveness or safety of the above uses. Also, sometimes the deadly nightshade, which resemble the burdock, can cause inadvertent harm if not removed from burdock harvests.

CALENDULA

Calendula officinalis L.
(family Asteraceae)

Description: Calendula is a common cultivated ornamental with orange or yellow flowers, also referred to as pot marigold or garden marigold. It should not be confused with members of the genus *Tagetes*, which are also widely grown and commonly called marigolds.

Range: Widely grown in gardens in the United States and Europe.

Application: During its long history, calendula has been administered internally for a variety of ailments, including spasms, fevers, suppressed menstruation, and cancer. Its chief use, however, was to help heal and prevent infection of lacerated wounds. Modern herbalists recommend it to heal a variety of skin conditions ranging from chapped hands to open wounds.

Comments: Research has begun to provide some scientific evidence to confirm calendula's long–standing traditional use in wound healing, although studies do not reveal any principles that are unique or outstanding in their physiological properties.

CHICORY

Cichorium intybus L.
(family Asteraceae)

Description: Anywhere from three to five feet in height, its attractive, conspicuous blossoms are azure blue, and its roots resemble those of dandelions.
Range: Native to Europe; naturalized in North America.

Application: Folk medicine and scant contemporary studies have shown chicory may function as a diuretic, laxative, remedy for tachycardia (rapid heartbeat), and as a poultice against inflammation. Its main use, however, is as a green for salads and as a substitute for coffee when its roots are dried and pulverized.

Comments: it has no more effect on the human nervous system and heart rate than the average cup of coffee.

ECHINACEA

Echinacea angustifolia DC.
(family Asteraceae)

Description: A perennial member of the daisy family, also known as cone flower, or narrow–leaved or purple cone flower. The plant has narrow leaves and a stout stem, up to three feet in height, that terminates in a single, large purplish flower head.

Range: Native to the central United States, where nine different varieties grow, the cultivated garden variety being the most commonly utilized.

Application: Originally used as an anti–infective, echinacea is no longer valued in conventional medicine, but continues to be employed in folk medicine. Taken internally, it works to increase the body's resistance to certain types of infections and is useful in preventing and treating the common cold. Externally, it is useful in the treatment of hard–to–heal wounds.

Comments: Echinacea's best–substantiated activity is its immune–stimulant effect. However, considerable controversy remains as to its relative effectiveness following oral administration, with capsules probably being less active than hydroalcoholic extracts. Significant side effects have not been reported, but allergic reactions to the plant are possible.

EVENING PRIMROSE

Oenothera biennis L.
(family Onagraceae)

Description: One of a family of the over 200 species of annuals, perennials, and biennials whose white, yellow, and reddish flowers are cultivated for their ornamental qualities.

Range: Temperate regions of the Western Hemisphere.

Application: Initial tests of primrose seed oil have indicated that it may be of some use for weight loss, lowering blood cholesterol, lowering blood pressure, helping rheumatoid arthritis, relieving premenstrual pain, retarding multiple sclerosis, alleviating hangovers, curbing mastalgia (sore breasts), preventing cardiovascular disease, and treating atopic eczema.

Comments: Evening primrose cannot be recommended for any of the above treatments until further research is conducted.

FENUGREEK

Trigonella foenum–graecum L.
(family Fabaceae)

Description: A small annual herb up to 20 inches tall, producing yellowish white flowers.

Range: Southern Europe and southwest Asia.

Application: The dried, ripe seeds have been used as a folk remedy for various ailments, including diabetes, hyperlipidemia, cellulitis, boils, tuberculosis, baldness, ulcers, and fungal infections. It was listed as the principal ingredient in Lydia Pinkham's Vegetable Compound. Also used as a spice and flavoring, since the seeds have a flavor somewhat similar to maple syrup.

Comments: Although not a particularly potent medicament, it is quite harmless in normal use.

FEVERFEW

Tanacetum parthenium (L.) Shultz Bip.
(family Asteraceae)

Description: A strongly aromatic perennial herb that is widely cultivated as an ornamental.

Range: Probably native to southeast Europe.

Application: Since the time of Dioscorides (78 A.D.), feverfew has been used for the treatment of headache, menstrual irregularities, stomachache, and especially fevers.

Comments: Considerable evidence has been obtained to confirm feverfew's effectiveness in reducing the frequency and severity of migraine attacks by using whole leaf preparations. Feverfew is contraindicated in individuals with hypersensitivity to other plants in the daisy family, and should not be used by pregnant or lactating women.

LICORICE

Glycyrrhiza glabra L.
(family Fabaceae)

Description: Consists of the dried rhizome and roots of a leguminous plant with pinnate leaves and spikes of blue flowers.

Range: Eastern Mediterranean region.

Application: The root has long been used as a flavoring in tobacco products, as an ingredient in pharmaceuticals, especially throat lozenges, in making candy, and for its expectorant and demulcent properties in the treatment of coughs and colds.

Comments: Although licorice does have a pleasing flavor and may have some utility in treating coughs and other conditions, it is a potent botanical. Large doses over extended periods of time are toxic.

ST. JOHN'S WORT

Hypericum perforatum L.
(family Hypericaceae)

Description: An aromatic perennial herb, producing perfectly balanced five–petaled golden yellow flowers and untoothed leaves usually marked with blackish spots, which are oil–bearing glands.

Range: Native to Europe, but prevalent as a weed throughout the United States and subtropical and temperate regions worldwide.

Application: Medical use of St. John's wort goes back as far as Dioscorides and Hippocrates. Today, it is used internally to treat mild to moderate depression, and externally to treat inflammatory problems such as hemorrhoids.

Comments: The hypericin in St. John's wort can lead to dermatitis and inflammation of the mucous membranes when exposed to direct sunlight, but this rarely, if ever, occurs with normal doses. Extended use of St. John's wort should be discontinued if such symptoms persist.

SASSAFRAS

Sassafras albidum (Nutt.) Nees
(family Lauraceae)

Description: A shrub or small tree with ovoid and deeply lobed leaves.

Range: Eastern North America.

Application: A tea prepared from the root bark has enjoyed a considerable reputation as a stimulant, antispasmodic, sudorific (sweat producer), depurative ("purifier") and as a treatment for a variety of diseases.

Comments: Despite its pleasant flavor and folkloric reputation as a useful tonic the plant material has no really significant medical or therapeutic utility. The volatile oil obtained from sassafras contains about 80 percent safrole, recognized as a carcinogenic agent in laboratory animals.

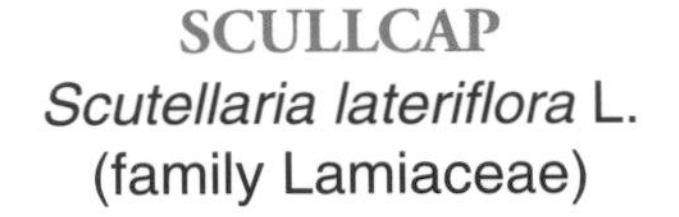

SCULLCAP

Scutellaria lateriflora L.
(family Lamiaceae)

Description: An erect perennial reaching a height of about two feet with two–lipped, brightly colored flowers.

Range: Grows in wet places in Canada and the northern and eastern United States.

Application: Scullcap has long been reputed to have tonic, tranquilizing, and antispasmodic effects. As recently as 1990 it was praised for its supposed beneficial effects on the nervous system. In the past it was a common ingredient in proprietary remedies for “female weakness.”

Comments: Research dating as far back as 1916 has shown that scullcap has no medicinal value. However, an herb called germander (*Teucrium*), which can damage the liver in excessive doses, is sometimes mistakenly sold as scullcap, making it even more important to avoid this herb.

TEA TREE

Melaleuca alternifolia
(Maiden and Betche) Cheel
(family Myrtaceae)

Description: A tall, bushy shrub with small, leathery, and highly aromatic evergreen leaves.

Range: Native to Australia.

Application: The volatile oil is obtained from the leaves by steam distillation and is a very popular home remedy in Australia for the treatment of cuts, abrasions, burns, insect bites, and athlete’s foot.

Comments: Limited clinical studies indicate tea tree oil’s possible value as an antimicrobial agent in the treatment of various vaginal and skin irritations. Its topical use is not generally associated with any toxicity.

Steven Foster
Varro E. Tyler, PhD, ScD

Tyler's Honest Herbal
A Sensible Guide to the Use of Herbs and Related Remedies

Fourth Edition

REVIEWS, COMMENTARIES, EVALUATIONS . . .

"Four stars! This fourth edition is a welcome addition to the references available on herbal products. The authors are well-recognized authorities in this field and they use their expertise to give the reader an unbiased evaluation of commonly used herbals. This reference can be of use to both health care professionals and consumers as a commonsense guide to herbal products."

Doody's Electronic Journal

"The fourth edition of *Tyler's Honest Herbal* provides essential botanical information as well as folkloric background of herbal remedies in a clear, accessible style. Unlike other herb books, this book gives you a serious evaluation of both the positive and negative features of the use of the most important herbs for therapeutic purposes. This new edition features additional scientific data on herbs that will enable you to make wise, informed choices about the benefits and risks of herbs currently available."

Herb Federation of New Zealand Journal

More REVIEWS, COMMENTARIES, EVALUATIONS . . .

"No hype, just the facts. This book is what I look at first when considering herbal therapy."

Isadore Rosenfeld, MD
Ida and Theo Rosai Distinguished Professor of Clinical Medicine; Attending Physician, The New York Hospital-Cornell Medical Center

"This is a very important, appropriately named book! Health practitioners and the lay public alike need *Tyler's Honest Herbal*, especially now that interest in herbal medicine is at an all-time high. Unfortunately, herbs that are good, worthless, and even bad can be found on the same store shelf. Separating them is what this book is all about, telling the differences clearly and objectively in a compact, factual guide that stands virtually alone amid the massive marketing efforts of a thousand herb sellers. The authors deserve special kudos for their extensive use of current, peer-reviewed scientific articles as a basis for making conclusions.

In November 1998, when the *Journal of the American Medical Association* opened the door for herbs to enter the realm of evidence-based treatment, the editors said, 'In God we trust. All others must have data.' Now Steven Foster and Dr. V. E. Tyler—the acknowledged dean of herbal scientists—follow with their compendium of the available herbal data, untainted by commercialism and refined into a highly readable fourth edition of this bible. The timing couldn't be better. *Tyler's Honest Herbal* will quickly become a trusted resource for anyone contemplating prescribing or taking an herbal therapy. And for honest herb sellers everywhere, the presence of this book alongside a product display will help to distinguish their wheat from the chaff of many opportunistic competitors."

Leonard S. Marks, MD
Medical Director, Urological Sciences Research Foundation; Clinical Associate Professor, Department of Urology, UCLA School of Medicine

"An up-to-date revision of the most reliable source for the layperson on herbal medicines. Provides a clear, concise summary of some of the most important and popular herbs and related products. Written in Tyler's usual smooth and succinct style. Excellent as a starting point for scientists who desire more information on herbal medicines."

Norman R. Farnsworth, PhD
Research Professor of Pharmacognosy, College of Pharmacy, University of Illinois, Chicago

REVIEWS, COMMENTARIES, EVALUATIONS . . .

"This is a very important, appropriately named book! Health practitioners and the lay public alike need *Tyler's Honest Herbal,* especially now that interest in herbal medicine is at an all-time high. Unfortunately, herbs that are good, worthless, and even bad can be found on the same store shelf. Separating them is what this book is all about, telling the differences clearly and objectively in a compact, factual guide that stands virtually alone amid the massive marketing efforts of a thousand herb sellers. The authors deserve special kudos for their extensive use of current, peer-reviewed scientific articles as a basis for making conclusions.

In November 1998, when the *Journal of the American Medical Association* opened the door for herbs to enter the realm of evidence-based treatment, the editors said, 'In God we trust. All others must have data.' Now Steven Foster and Dr. V. E. Tyler—the acknowledged dean of herbal scientists—follow with their compendium of the available herbal data, untainted by commercialism and refined into a highly readable fourth edition of this bible. The timing couldn't be better. *Tyler's Honest Herbal* will quickly become a trusted resource for anyone contemplating prescribing or taking an herbal therapy. And for honest herb sellers everywhere, the presence of this book alongside a product display will help to distinguish their wheat from the chaff of many opportunistic competitors."

Leonard S. Marks, MD
Medical Director, Urological Sciences Research Foundation; Clinical Associate Professor, Department of Urology, UCLA School of Medicine

The Haworth Herbal Press
An Imprint of The Haworth Press, Inc.

NOTES FOR PROFESSIONAL LIBRARIANS AND LIBRARY USERS

This is an original book title published by The Haworth Herbal Press, an imprint of The Haworth Press, Inc. Unless otherwise noted in specific chapters with attribution, materials in this book have not been previously published elsewhere in any format or language.

CONSERVATION AND PRESERVATION NOTES

All books published by The Haworth Press, Inc. and its imprints are printed on certified pH neutral, acid free book grade paper. This paper meets the minimum requirements of American National Standard for Information Sciences-Permanence of Paper for Printed Material, ANSI Z39.48-1984.

Tyler's Honest Herbal

A Sensible Guide to the Use of Herbs and Related Remedies

Fourth Edition

THE HAWORTH PRESS
Titles of Related Interest

Tyler's Herbs of Choice: The Therapeutic Use of Phytomedicinals, Second Edition by James E. Robbers and Varro E. Tyler

Tyler's Honest Herbal: A Sensible Guide to the Use of Herbs and Related Remedies, Fourth Edition by Steven Foster and Varro E. Tyler

Herbal Medicinals: A Clinician's Guide by Lucinda Miller and Wallace J. Murray

Drugs of Natural Origin: Economic and Policy Aspects of Discovery, Development, and Marketing by Anthony Artuso

Access to Experimental Drugs in Terminal Illness: Ethical Issues by Udo Schüklenk

Opium Poppy: Botany, Chemistry, and Pharmacology by L. D. Kapoor

Dictionary of Plant Genetics and Molecular Biology by Gurbachan S. Miglani

Horticulture as Therapy: Principles and Practice by Sharon P. Simson and Martha C. Straus

Plant Alkaloids: A Guide to Their Discovery and Distribution by Robert F. Raffauf

Seed Quality: Basic Mechanisms and Agricultural Implications edited by Amarjit S. Basra

People-Plant Relationships: Setting Research Priorities edited by Joel Flagler and Raymond P. Poincelot

Glossary of Vital Terms for the Home Gardener by Robert E. Gough

Herbs, Spices, and Medicinal Plants, Volumes 1, 2, 3, and 4 edited by Lyle E. Craker and James E. Simon

Tyler's Honest Herbal

A Sensible Guide to the Use of Herbs and Related Remedies

Fourth Edition

Steven Foster
Varro E. Tyler, PhD, ScD

The Haworth Herbal Press
An Imprint of The Haworth Press, Inc.
New York • London

First published 1982, *The Honest Herbal*
Second edition 1987, *The New Honest Herbal*
Third edition 1993, *The Honest Herbal*
Fourth edition 1999, *Tyler's Honest Herbal*

Softcover edition published 2000.

Published by

The Haworth Herbal Press, an imprint of The Haworth Press, Inc., 10 Alice Street, Binghamton, NY 13904-1580

Cover design by Jennifer M. Gaska.

The Library of Congress has cataloged the hardcover edition of this book as:

Foster, Steven, 1957-
Tyler's honest herbal : a sensible guide to the use of herbs and related remedies / Steven Foster, Varro E. Tyler.—4th ed.
p. cm.
Rev. ed. of: The honest herbal / Varro E. Tyler, 1992.
Includes bibliographical references and index.
ISBN 0-7890-0705-3 (alk. paper).
1. Herbs—Therapeutic use. 2. Materia medica, Vegetable. I. Tyler, Varro E. II. Tyler, Varro E. Honest herbal. III. Title.
RM666.H33T94 1998
615'.321—dc21 98-43234
CIP

ISBN 0-7890-0875-0 (pbk.)

To Ginny, of course

ABOUT THE AUTHORS

Steven Foster is President of Steven Foster Group, Inc., Fayetteville, Arkansas. Foster is a well-known author, photographer, and consultant in the field of medicinal and aromatic plants. He began his career nearly a quarter century ago at the Herb Department of the Sabbathday Lake, Maine, Shaker Community. Foster is the author of ten books on herbs, their cultivation, and use, including *A Field Guide to Medicinal Plants: Eastern and Central North America* (with James A. Duke), *Herbs for Your Health,* and junior author of *Encyclopedia of Common Natural Ingredients Used in Food, Drugs, and Cosmetics* (with Albert Y. Leung), Second Edition. Foster has authored more than 600 articles in popular, trade, and scientific publications and serves as Senior Editor of Herbs and Medicinal Plants—A Multidisciplinary Approach, a book program that is part of The Haworth Herbal Press.

Varro E. Tyler, PhD, ScD, Editor in Chief of The Haworth Herbal Press, is Distinguished Professor Emeritus of Pharmacognosy in the School of Pharmacy and Pharmacal Sciences at Purdue University. A recognized authority on plant drugs (herbs) and their uses, he is a co-author of *Pharmacognosy and Pharmacobiotechnology,* the standard U.S. textbook in the field. Dr. Tyler is the author of over 300 scientific and educational publications, including more than two dozen books. He is a member of the editorial boards of several journals and a Fellow of the Academy of Pharmaceutical Research and Science, the Academy of Pharmaceutical Scientists, and the American Association for the Advancement of Science. Dr. Tyler served as Purdue University's executive vice president for academic affairs for five years, and for twenty years as Dean of its School of Pharmacy and Pharmacal Sciences.

CONTENTS

Preface

Medicine and quackery have always been close, if not compatible, partners. At times, they may appear to have separated, but sooner or later, in one place or another, they wind up reunited. The present area of greatest mutual attraction for them appears to be in the treatment of diseases by means of herbal remedies. More misinformation regarding the efficacy of herbs is currently being placed before consumers than at any previous time, including the turn-of-the-century heyday of patent medicines.

I had long wanted to write a popular herbal because, during three decades as a professional pharmacognosist, I had often wished to use this means to convey to others some of my interest in and enthusiasm for these fascinating natural drugs. But recently, after being exposed to some of the deluge of inaccurate and deceptive information which has appeared on the subject, I felt impelled to write about it. The result is this book.

The bulk of the useful information is contained in the write-ups on individual drugs which follow the introductory sections. Most of them are organized according to a similar plan: a brief description of the drug and its proper nomenclature, as well as that of the plant from which it is derived, precede comments on its alleged uses. Then follows a nontechnical discussion of the chemistry and pharmacology (where known) of the active principles of the drug. Next is an evaluation based on all evidence known to the author which offers a judgment about the probable utility of the herb. Pertinent references to the literature conclude this discussion. These provide documentation for all significant statements and are possibly a unique feature of a book designed specifically for everyone interested in herbs. However, the references must be scrutinized carefully; some represent uncritical advocacy literature, others are scientific and authoritative. The context where each is cited will immediately clarify the type of literature involved.

Some readers will probably find fault with the information in this book, especially with the judgments presented. Some will find it too conservative, and it is, compared to the writings of most modern herbalists. Others may believe the volume is too liberal, promoting self-medication with self-selected herbs. That is a misconception. Self-medication is *not* being promoted. The information given only permits it to be done as intelligently as possible, if the reader wants to do so in the first place.

The lay reader, for the first time, is thus provided with accurate, scientific statements and reasoned judgments based on a recognized authority's lifetime of professional study of medicinal plants. Basic facts on the uses of herbs and related remedies are greatly needed, for interest in the entire field is tremendous and growing.

Because of this interest, it was impossible to cover all of the herbs of supposed medicinal value in a book of manageable size. One of the older listings of botanical drugs, *J. M. Nickell's Botanical Ready Reference,* first published a century ago, named 2,526 different medicinal plants ranging from *Abelmoschus esculentus* (L.) Moench, the common okra, to *Zizia aurea* (L.) W. D. J. Koch, otherwise known as the meadow parsnip. The coverage could be extended even more if based on introductions since that time. Physical limitations aside, it is really not necessary to discuss all of these botanicals, for many are seldom encountered or are totally without value. The more than 100 included here were selected on the basis of relative significance to the public determined from literature sources (including retail and wholesale catalogs of herbs) and actual observation of products sold in health food stores.

There are three ways in which herbs are ordinarily obtained. The most common, in this era of urban living, is by purchase in a retail store. The problems in verifying the true identity of such products are discussed later. Some consumers may either choose to grow their own or to collect them from the wild. These options are completely beyond the scope of this book. While many recent herbal writings do encourage both activities, a word of warning is needed, especially in collecting wild plants. Make absolutely certain of the identity of the harvested material. Natural variations in different specimens can be misleading, and thus plant taxonomy (i.e., identification) is neither an exact science nor an easy one. Fatalities have been reported as a result of confusing and subsequently ingesting one wild plant for another.

Whether plant drugs are purchased or self-collected, it is important to remember that their active constituents may vary considerably, depending on: the conditions under which the plant was grown, the degree of maturity at the time of collection, the manner of drying, the conditions of storage, and other similar factors. These variables are overcome, in the case of crude drugs used in conventional medicine, by conducting chemical or physiological assays or tests and then standardizing the product by adding drug material of greater or lesser potency. The more potent the medicine, the more important it is that some sort of control be utilized to assure proper dosage. Unfortunately, herbs are seldom subjected to such procedures.

The information on the various herbs has been arranged in alphabetical order by common name for the convenience of the reader. Today, few people are able to place a particular plant in its proper family, let alone various families in their proper phylogenetic order. Imprecise as they are, common names have been used as titles simply because they are ones most people know. The imprecision in the system, I hope, has been overcome by comprehensive indexing.

Finally, remember as you read that your interest in herbs should be constructive, not destructive, to your health. This requires an ability not only to seek the truth but, after finding it, to discard any preconceived ideas which it may reveal as untrue.

> Beware of the truth, gentle Sister. Although much sought after, truth can be dangerous to the seeker. Myths and reassuring lies are much easier to find and believe. If you find a truth, even a temporary one, it can demand that you make painful changes.*

V. E. T.
West Lafayette, Indiana
December 1981

---

*Excerpted with permission of the author from *God Emperor of Dune*. Copyright © 1981 by Frank Herbert.

Preface
to *The New Honest Herbal*

Much has happened in the field of herbal medicine since the manuscript for *The Honest Herbal* was completed more than five years ago. As predicted, some herbal enthusiasts found the information in that book too conservative. It nevertheless received uniformly excellent critical acclaim from knowledgeable reviewers and readers; the original edition sold out completely.

In the interim, there have been some very encouraging developments in the field, stimulated, I believe, in no small measure by *The Honest Herbal.* Several consumer-oriented groups, such as the National Council Against Health Fraud and the American Council on Science and Health, have begun to inform the public about the true utility of herbs. Several popular monthly newsletters, such as *Nutrition Forum* and *Tufts University Diet and Nutrition Letter,* have begun to feature articles disclosing the real facts about herbal products. It is significant to note, however, that these publications have as their primary focus the field of nutrition, not pharmacy or medicine. Many American journals and newsletters in these latter two areas still disdain authoritative articles dealing with herbs and their uses.

Consumer Reports, an extremely influential nationally circulated magazine devoted to examining product quality, has recently featured exposés of malpractice in the herbal field. And a very hopeful sign is the editorial reorientation of *Prevention,* a popular health magazine with 2.75 million circulation, toward a more conservative and genuinely helpful stance with regard to the utility of herbal products.

Events of this sort provided a considerable incentive to prepare a thorough revision of my original work. Consequently, the entire book has been revised and new material added, where appropriate, to produce *The New Honest Herbal.* It discusses encouraging new developments in the field of herbal regulation and control taking place in Canada. It contains new monographs on several important herbs, including butcher's-broom, capsicum, evening primrose, feverfew, ginger, pau d'arco, and schisandra. Some of the monographs, including those on aloe, hops, and yohimbe, contain important additions. All of the monographs have been carefully reviewed, and many are updated with significant new information and references.

The result is a volume of even greater significance than my first effort to those interested in herbs. It is now made available in softcover format so that it can reach a wider audience than its predecessor. I hope that all who read *The New Honest Herbal* will benefit from the uniquely honest and useful herbal information found within its pages.

V. E. T.
West Lafayette, Indiana
January 1, 1987

Preface
to *The Honest Herbal*
Third Edition

Once again, an edition of this book has been completely sold out and revision is required. In the years since *The New Honest Herbal* was published, substantial advances have occurred in some areas of herbal medicine, and these are incorporated into the appropriate sections and monographs in this third edition. Unfortunately, most of the progress has taken place in countries outside the United States where the laws and regulations pertaining to the marketing of these time-honored remedies are more realistic. One positive development here has been the maturation of *HerbalGram,* a publication of the American Botanical Council that has become an excellent source of reliable herbal information, even if most of the new research reported there occurs in other countries.

Changes in this edition include the addition of six new monographs on commercially significant herbs and herbal products: cranberries, ephedra, ginkgo, milk thistle, suma, and tea tree. Some write-ups, such as the one on L-tryptophan, have been substantially revised to reflect the dramatic changes that have taken place in its use. Others, such as buchu, are essentially unchanged, indicating the lack of attention this herb has received in the interim.

Changes in many of the classical scientific names of plant species and even the families to which they belong have been made, albeit reluctantly, to bring them in line with modern botanical thought. While it was considered necessary to do this, it does present a confusing situation to most readers. At least five different scientific names for German chamomile are now found in the literature, and the "correctness" of any of them is, to some extent, a matter of opinion. The names utilized in this work have been chosen on the basis of information that appears to be the most authoritative at this time.

One feature that continues to be emphasized in this edition, as it was in the two previous ones, is that herbs are actually nothing more than diluted drugs. Regardless of the regulatory or legal advantages that may be gained by calling them something else, whenever herbs are used to diagnose, cure, mitigate, treat, or prevent disease, they are, by definition, drugs.

They do not possess any magical or mystical properties, and like other drugs, they must be administered in proper doses for appropriate periods of time to produce their benefits. These benefits result from the presence of one or more active principles, usually complex chemical compounds, that are present in the plant material. Every herb is different from every other herb. Some are safe and effective. Some are neither. As is the case with other drugs, the administration of herbs may produce undesirable side effects.

It is necessary to understand all of these facts if one is to utilize herbs safely and effectively for therapeutic purposes. *The Honest Herbal,* Third Edition, is one of the few books available in any language that provides this kind of information to the reader. It shuns hyperbole and unsubstantiated claims. It presents the necessary documentation of significant statements to allow its readers to consult the original literature and determine for themselves the proper context and interpretation of the data. If it errs, it errs on the side of conservatism because its author believes that drugs of any kind, natural or synthetic, prescription or self-selected, should improve the consumer's health, not cause it to deteriorate. That is what herbal medicine—or any kind of medicine—is all about.

This work could not have been written without the assistance of several people. I wish to express sincere gratitude to my wife, Ginny Tyler, who prepared the index and who facilitated the preparation of the book in every way possible. Appreciation is also extended to Theodora Andrews, Professor of Library Science, who diligently searched the literature for appropriate references and to my assistant, Linda Michael, who patiently typed, retyped, and computerized the numerous revisions. Thank you!

V. E. T.
West Lafayette, Indiana
1993

Preface
to *Tyler's Honest Herbal*
Fourth Edition

Nearly sixteen years have passed since publication of the first edition of *The Honest Herbal*, a work that has become one of the most often referenced and quoted modern herbals. It has also become one of the most controversial modern herbals, often evoking emotional reactions from herbal advocates who do not wish to see their favorite herbs verbally abused under the guise of science. However, for herbs to reach their rightful role in making a significant contribution to affordable health care in the United States, they must be assessed from a scientific viewpoint that ensures: (1) proper botanical identity of plant material; (2) predictable chemical consistency from batch to batch; (3) predictable therapeutic benefits, based not only on traditional knowledge but also on well-designed, controlled clinical studies; and (4) proof of safety. The various editions of *The Honest Herbal* have reflected major and significant scientific advancements in herbal research. When convincing, that data has been presented in a favorable light.

Each edition has also reflected how herbs are perceived in the modern marketplace. Since the early 1980s, that market has increased beyond anyone's wildest expectations. In the ensuing six years since publication of *The Honest Herbal*, Third Edition, in 1993, the American herb market has changed more than at any comparable point in history. Herb products intended for health benefits were once relegated to health and natural food retail outlets in the United States. The number of health food stores has nearly tripled since this book was first published. Since the passage of the Dietary Supplement Health and Education Act of 1994, herb products have found their way into every conceivable retail outlet where nonprescription drugs and personal care products are sold, including independent pharmacies, chain pharmacies, discount department stores, supermarkets, and even checkout counter displays at convenience stores. Sales at the retail level have jumped from an estimated $500 million in 1992 to projections of over $5 billion in sales for 1999. Growth of the herbal dietary supplement industry continues at over 30 percent per year. The faces of Wall Street investors are becoming as commonplace as those of scientists at the plethora of herbal conferences and symposia held each year.

The entry of herbs into the mass market has created a "cart before the horse" situation for education of health care professionals in the retail environment, particularly for pharmacists. Buyers for chain drugstores or discount department store pharmacies have filled shelf space with herb products before a new generation of pharmacists has had an opportunity to learn of the risks, benefits, proper dosage forms, contraindications, and side effects of the thousands of herb products that have flooded the market. Call them dietary supplements, call them fairy dust, herbs are still—by any reasonable definition—drugs. A pharmacist selling herb products or advising consumers on product benefits should know that milk thistle *fruit* (commonly called seed) extracts standardized to 80 percent silymarin may be beneficial for certain liver conditions. He or she should also know that there is not one shred of scientific evidence to suggest that therapeutic benefit can be expected from powdered milk thistle *leaf* products in whatever form they are sold.

Fortunately, several organizations have begun to fill this information void for professionals. Tyler and Foster contributed a chapter on herbs and phytomedicinals to the eleventh edition of the *Handbook of Nonprescription Drugs* published by the American Pharmaceutical Association (1996). The Austin, Texas–based, American Botanical Council is offering continuing education programs on herbs in conjunction with the Texas Pharmacy Foundation. Major schools of pharmacy at Purdue University, the University of Illinois at Chicago, and University of Mississippi, to name a few, have also instituted pharmacy continuing education programs on phytomedicines.

In recent decades, Dr. Norman R. Farnsworth of the Program for Collaborative Research in the Pharmaceutical Sciences, University of Illinois at Chicago, stamped his correspondence with the words

"Save the Endangered Species: Pharmacognosy." That academic discipline, dealing with various multidisciplinary aspects of medicines derived from natural sources, is experiencing a slow rebirth. Until recently, most graduate students entering such programs were foreign students. Now, American students are again following academic career paths toward a pharmacognosy degree. Within the next five years, all medical schools in the United States will offer at least undergraduate programs in so-called alternative or complementary medicine. Still, professional education has been slow to catch up with the explosive marketplace.

Tyler's Honest Herbal, Fourth Edition, is intended to bring scientific understanding of commonly sold herbs into the twenty-first century. Again, if we err, we do so on the side of conservatism. The current interest in herbs is consumer driven. Now that herb products are much more widely available, offered more often than not repackaged by marketing companies with no internal scientific knowledge of the substances they purvey, truth can indeed be stranger than fiction.

Steven Foster
Fayetteville, Arkansas

Pros and Cons

The resurgence of interest in herbal medicine that originated during the last decade shows every promise of continuing its rapid development into the next century. Apparently, the movement had its origin in many people's disillusionment with modern medicine—its high cost and its inability to cure everything. This, along with the widespread belief that plant remedies are "naturally" superior to man-made drugs, produced a wave of enthusiasm and promotion on the part of the public that can only be described as an herbal renaissance.

Unfortunately, many partisans in the revival were more enthusiastic than knowledgeable, more evangelistic than critical. For reasons that will be discussed in detail, herbs could be sold legally if they were not labeled for use in the treatment of disease. What this means is that advocacy literature containing the most outrageous claims of therapeutic effectiveness could be sold side by side with the drugs, so long as it was not part of the official labeling. To protect the author, most such books and pamphlets carry a carefully worded disclaimer indicating that the reader should refrain from testing any of the suggested remedies but, instead, should consult a physician on all matters pertaining to drugs and therapy. The consumer is thus taunted with information he or she is warned not to apply!

Voluminous is too conservative a word to apply to the existing quantities of this kind of promotional literature. Staggering is a better term. These modern herbals range from small, cheaply printed, paper-covered pamphlets dealing with single or, at most, small groups of drugs to large, elaborately produced, comprehensive studies in fine bindings with attractive artwork and numerous color plates. Some combine herbal lore with astrology. Others, in an obvious attempt to cultivate the interest of drug abusers, emphasize plant substances with mind-altering properties. Still others slant their coverage toward certain groups such as women, vegetarians, youth, the aged, or the outdoorsman. Another type of promotional literature may be viewed as developing from the now discredited "American school of eclectic medicine," in which special attention was given to plant remedies. This has been characterized as "the apotheosis of the old grandmother and witch-doctor systems of treatment."[1]

Now add the Internet to the information milieu. In essence, the Internet provides the opportunity for a medicine show in every home, on every desk with a computer and modem. Search for information on a single herb, and you may retrieve thousands of Web links, the vast majority of which offer promotional literature for products. Even more hyperbole is possible in this worldwide wasteland of information, where anyone can "publish" what they like without the benefit of any editorial screening. Of course, the World Wide Web can also be a valuable tool. MedLine, the database of the National Library of Medicine in Washington, DC, is freely available to all who wish to search, along with dozens of authoritative informational databases on botany, chemistry, pharmacy, and medicine. One simply has to learn to discern between fact and fiction.

Practically all nonscientific advocacy literature recommends large numbers of herbs for the treatment of a variety of ailments based on hearsay, folklore, and tradition; in fact, the only criterion that seems to be rigorously avoided is scientific evidence. Some are so comprehensive and indiscriminate that they appear to recommend everything for anything. Although such lack of judgment should be deplored, it is not nearly so potentially harmful as the many instances in which downright dangerous, even deadly, poisonous herbs are recommended, usually on the basis of some outdated reference or a misunderstanding of the facts.

Particularly insidious is the myth perpetrated by these promoters that there is something almost *magical* about herbal drugs which prevents them—in their natural state—from inflicting harm on living organisms. Think how completely false this argument is! Even those unfamiliar with the execution of Socrates by poison hemlock more than two thousand years ago are probably not inclined to collect and eat wild mushrooms indiscriminately.

To understand plant drugs completely, we need to know their botany, chemistry, and pharmacology. Few modern herbalists possess such a comprehensive background. Therefore, they rely heavily on outdated writings and, without benefit of modern scholarship, transmit the opinions and recommendations of authors who long ago ceased to be authorities. We do not take our automobiles to the livery stable to be serviced; neither should we depend on the sixteenth-century herbalist John Gerard or the seventeenth-century apothecary-astrologer Nicholas Culpeper for modern therapeutic advice. Culpeper's writings are full of astrological explanations for the efficacy of various drugs (which may explain his renewed popularity in the "Age of Aquarius").

However, recent commentators on Culpeper's writings seem to know even less than he did. One of them tells us that lily-of-the-valley (*Convallaria majalis* L.) ". . . not being poisonous, does not leave any harmful results if it is taken over a long period."[2] Yet all modern authorities characterize the plant as poisonous, and convallatoxin, the principal glycoside contained in it, is regarded as the most toxic cardiac glycoside in existence.[3]

Similar misstatements are not confined to commentaries on Culpeper. Aikman's attractive compilation of folk medicine features photographs of a Virginia woman brewing up an oversized kettle of comfrey leaves to make "a sweet tea for calming coughs and stomach ulcers."[4] This, despite evidence that comfrey contains several hepatotoxic pyrrolizidine alkaloids shown to produce cancer of the liver in small animals.[5] Gibbons gives us a recipe for coltsfoot cough drops and says he drinks a medicinal tea made from that plant for "pure pleasure."[6] Coltsfoot also contains a carcinogenic pyrrolizidine alkaloid, senkirkine.[7] Writing in 1977, A. and S. McPherson recommended iced sassafras tea as a "great summer drink" and even gave us a recipe for sassafras jelly.[8] Yet, since 1960, sassafras oil and its major constituent, safrole, have been banned as flavoring agents because of their carcinogenicity.[9] A large number of therapeutic uses for pokeroot are described by Santillo[10] with no mention of the fact that it is so poisonous that persons drinking a single cup of tea made from it have required hospitalization.[11] Candy lovers eating large amounts of licorice have also become seriously ill,[12] but F. and V. Mitton tell us it has "the greatest value in cough medicine" without a word of warning.[13] The list could go on and on. Enough examples have been cited to demonstrate the lack of critical evaluation in the current herbal literature—and the wisdom of reasoned skepticism.

This brings us to a discussion of another of the tenets of modern herbalism: the dogma that whole drugs—that is, leaves or roots or seeds or the like—have physiological properties different from the active constituents isolated from the same plant parts. This, of course, is a fallacy. Although certain plants do contain a large number of active principles, and the observed effects following administration will be a combination of all of them, these various principles usually display activities that are qualitatively similar, if not quantitatively identical.

For example, digitalis, the dried leaf of *Digitalis purpurea* L., has long been used in the treatment of congestive heart failure. It contains approximately thirty different glycosides that possess some cardiotonic properties. One of them, digitoxin, is isolated from the leaf and marketed individually as a commercially important drug. Both products, digitalis and digitoxin are used in the same way for treatment of the same disease, namely, congestive heart failure. They differ in potency, time of onset, and in duration of activity, but they affect the heart in the same way and are therefore qualitatively similar.

Countless other examples could be given. Opium and its principal alkaloid, morphine, produce similar physiological effects. So do cinchona bark and quinine, peppermint leaf and peppermint oil, coffee and caffeine, and wheat germ oil and vitamin E. To be sure, some natural products contain two or more principles with different activities, but these are ordinarily supplementary, not antagonistic. Cod liver oil contains appreciable quantities of both vitamins A and D. These vitamins can be isolated, either from this or other sources, and administered separately at the same dosage with precisely the same effect as that of the natural mixture in the oil. Again, because it is important, remember that *herbs and other drugs exhibit the same types of activity as do the active principles isolated from them.*

Another myth perpetrated by many promoters of herbal therapy is that "natural" products—organic chemicals synthesized in nature by metabolic processes in plants and animals—possess an innate superiority over the same product produced in a chemical laboratory. Such claims at times actually border on mysticism, but then herbs have long been linked by some with astrological phenomena.

Anyone with a knowledge of biology, biochemistry, or even history will recognize that this vitalistic doctrine, which draws a sharp distinction between inorganic and organic materials, is nothing new. It was completely discredited more than 150 years ago when the German chemist Friedrich Wöhler succeeded in producing the natural organic compound urea from a solution of inorganic ammonium cyanate. When writers show their ignorance of the significance of this event, the wise reader must question the validity of other aspects of their writing as well.

This ignorance still exists today: there is no difference in the vitamin C, for example, obtained from natural biosynthetic processes in rose hips, or by synthetic processes in the laboratory of a chemical manufacturer. The word "natural" applied to such materials identifies only a source and does not imply a degree of superiority or inferiority. However, on the labels of various drugs and vitamins, it does indicate that the consumer may expect to pay several times the normal price for such an item.

Another term much misused in recent writings on herbs is "organic." Applied to a wide variety of "health foods," it denotes that the product was grown under conditions utilizing only natural fertilizers, such as manure, and that no pesticides of any type were applied to it. In some ways, this is an extension of the vitalistic doctrine just discussed. Plant physiologists have known for years that plants have no mechanism for differentiating whether nutrients such as calcium, potassium, and nitrogen are derived from organic or inorganic sources, provided they are in a form the plant can assimilate. If anything, the organic forms are less readily available and may have to be acted upon by soil microorganisms before they can be utilized by the plant. Thus, if rapid, profuse growth is desired (as is normally the case), *inorganic or synthetic fertilizers that provide nutrients in a readily available form have a definite advantage.*

Pesticide residues may, of course, pose health hazards if present in sufficient concentrations. Fortunately, most can be removed by proper cleansing, and for those which cannot, appropriate limits of safety have been established. Since exactly the same problem exists with other elements in our environment that we consume every day in huge quantities—including air and water—there would seem to be much less cause for concern about residues on food products. Some experts now believe that carcinogenic hazards from current levels of pesticide residues are of minimal concern in comparison to the risks imposed by the carcinogens naturally present in foods and herbs.[14] Besides, the typical consumer has absolutely no way of knowing whether the item he or she has purchased is "organic" or not. All the consumer knows is that he or she has paid a premium for something that was probably misrepresented.

At this point, you may ask, "If so many of the herbal remedies have little or no value, or may even be dangerous to a person's health, why have they become so popular in recent years? Why do so many people, especially those who are unusually health conscious, continue to demand and use them?" The answer lies, at least partly, in the so-called *placebo effect.*

"Placebo" comes from the Latin "I will please" and refers to a drug that provides relief for the patient through mental processes rather than through any physiological effect on the disorder. The placebo effect is thus a physiological improvement brought about by a psychological mechanism. Mind over matter is a simpler way of putting it. A placebo does for you what you think it will do.

Scientific studies have shown that placebos actually work about one-third of the time.[15] They were found to be effective for the relief of severe postoperative wound pain, cough, drug-induced mood changes, angina pain, headache, seasickness, and the common cold in an average of 35 percent of these patients. More interesting was the observation that placebos work even better when the goal of therapy is some change in behavior (drowsiness, alertness), in subjective sensation (pain or discomfort), or in a response controlled by the endocrine glands or by the autonomic nervous system (blood pressure, acid stomach, asthmatic breathing). Since many of the conditions for which herbal treatment is commonly used fall into these categories, it is completely understandable why some people find them to be of some value, at least some of the time.

Discussing the placebo effect brings me to why the various herbal materials in homeopathic medicine are not mentioned in this book. Homeopathy is a system of medicine proposed by Samuel Hahnemann late in the eighteenth century which hypothesizes that disease is cured by remedies which produce symptoms in a healthy subject resembling the disease in question. However, and this is very important, the cure takes place only when the medicine is administered in such small amounts as to fail to produce the symptoms. Hahnemann based the latter principle on the never-proven assertion that in illness the body is enormously more sensitive to drugs than in health. It led to claims that doses as small as 0.000001 grain (approximately 0.000000065 gram) of a medicament could be effective in curing disease. Indeed, such dilutions were referred to as "high potencies."

Practitioners of conventional medicine scoffed at such wild claims, and they still do. One critic versified his objections:

> The homeopathic system, sir, just suits me to a tittle,
> It proves of physic, anyhow, you cannot take too little;
> If it be good in all complaints to take a dose so small,
> It surely must be better still, to take no dose at all.[16]

The interest in unconventional medicine developed during recent years has also extended to homeopathy. Many modern herbals quote homeopathic uses for various drugs. Thoughtful consideration tells us that, whatever it may be, the system of homeopathy has no relation to the effective use of drugs of any kind. Probably the best that can be said about this now discredited treatment is that it demonstrates the therapeutic value of the placebo effect. Although often confused, there is no relationship between homeopathy and herbal medicine (phytomedicine).

This book is designed to help consumers decide for themselves, on the basis of the most recent scientific evidence, whether an herb they have considered taking for a particular condition is (1) really potentially useful, (2) essentially without value (except for its possible placebo effect), or (3) potentially hazardous, for one reason or another. Hazards, however, should not be thought of simply in terms of toxicity. To some extent, all self-treatment with herbs or any other kind of medication is potentially hazardous because it may cause the patient to neglect conditions that could respond to timely professional therapy but if neglected, could result in serious health problems. Then, too, there is always the hazard to the patient's pocketbook.

For all of these reasons, you are less likely to receive value for money spent in the field of herbal medicine than in almost any other.

REFERENCES

1. M. Fishbein. *Fads and Quackery in Healing.* Blue Ribbon Books, New York, 1932, p. 31.
2. C. F. Leyel. *Culpeper's English Physician and Complete Herbal.* Wilshire Book Co., N. Hollywood, California, 1972, p. 72.
3. G. Baumgarten. *Die Herzwirksamen Glykoside.* VEB Georg Thieme, Leipzig, Germany, 1963, p. 67.
4. L. Aikman. *Nature's Healing Arts.* National Geographic Society, Washington, DC, 1977, pp. 30-31.
5. N. R. Farnsworth. *American Journal of Pharmaceutical Education* 43: 242, 1979.
6. E. Gibbons. *Stalking the Healthful Herbs,* Field Guide Edition. David McKay Co., New York, 1970, pp. 31-32.
7. I. Hirono, H. Mori, and C. C. J. Culvenor. *Gann* 67:125-129, 1976.
8. A. McPherson and S. McPherson. *Wild Food Plants of Indiana.* Indiana University Press, Bloomington, Indiana, 1977, pp. 99-101.
9. A. B. Segelman, F. P. Segelman, J. Karliner, and R. D. Sofia. *Journal of the American Medical Association* 236: 477, 1976.
10. H. Santillo. *Natural Healing with Herbs*, Eighth Printing. Hohm Press, Prescott, Arizona, 1990, pp. 162-163.
11. W. H. Lewis and P. R. Smith. *Journal of the American Medical Association* 242: 2759-2760, 1979.
12. J. W. Conn, D. R. Rovner, and E. L. Cohen. *Journal of the American Medical Association* 205: 492-496, 1968.
13. F. and V. Mitton. *Mitton's Practical Modern Herbal.* W. Foulsham and Co. Ltd., London, 1976, p. 104.
14. B. N. Ames, R. Magaw, and L. S. Gold. *Science* 236: 271-280, 1987.
15. M. C. Gerald: *American Pharmacy* NS19(5): 246, 1979.
16. *United States Magazine, and Democratic Review* 22: 418, 1848. In *Homeopathy in America,* M. Kaufman. The Johns Hopkins Press, Baltimore, Maryland, 1971, p. 30.

Laws and Regulations

Herbs abounded in old-time drugstores. Pharmacists used them mostly to prepare various kinds of solutions and extracts that were then mixed with other ingredients to fill prescriptions. During the past few decades, these drugs in their handsome glass-labeled bottles gradually disappeared from the shelves of pharmacies, replaced largely by packages of prefabricated medicines. Some shops retained a few, along with fancy glass globes of colored water, for show, not for use.

Part of the reason for this change is obvious. Pharmaceutical manufacturers could prepare medicines better, more accurately, and cheaper than any individual could. Not so obvious are the complicated laws and regulations that also exerted great influence on the disappearance of crude drugs from pharmacies. Not that they disappeared from commerce—far from it. They were simply moved to the so-called health food stores and now the mass market where they continue to exist in great variety and to be sold in enormous quantity. In 1990, the sale of herbs and herb teas in such establishments amounted to more than $467 million. The sale of books explaining the use of herbs and similar products totaled an additional $94 million in that year. In 1999, sales of herb products and support literature will exceed $5 billion. Let us look at the laws and regulations that helped to create this entire situation.

The original Food and Drugs Act of 1906 was, at the time, a bold step forward. It effectively abolished the patent medicine and meatpacking frauds that, thanks to sensational journalism, had been the main causes of the public pressure which led to eventual reform. The act prohibited adulterated or misbranded drugs but did not deal with the safety or efficacy of the drugs themselves. Those matters were really not addressed for a third of a century.

In the late 1930s, public opinion was again mobilized, this time by the Elixir-of-Sulfanilamide tragedy in which more than 100 persons were fatally poisoned by a newly marketed drug product.[1]

As a result, the 1938 Federal Food, Drug, and Cosmetic Act was passed. It required that all drugs sold in this country be *proven* safe. This act was subsequently amended in 1962. Since the Drug Amendments of 1962 followed extensive congressional investigations of the drug industry led by Senator Estes Kefauver, they are still commonly referred to as the Kefauver-Harris Amendments. They required that *all drugs marketed in the United States after 1962 be proven both safe and effective.*

The procedure used since 1938 to ensure new drug safety was known as a New Drug Application. That title soon became abbreviated to NDA. Something now had to be done so that those drugs already proven safe through the NDA mechanism were also proven effective. Approximately 4,000 different drug formulations, representing about 300 different chemical entities, fell into the category of drugs actually being sold, and another 3,000 formulations were covered by NDAs but were not actively marketed.

Lacking the resources to tackle this gigantic task itself, the Federal Food and Drug Administration (FDA) turned for help in 1962 to the Division of Medical Sciences of the National Academy of Sciences, National Research Council. They, in turn, organized a "Drug Efficacy Study," which lasted nearly seven years and finally culminated in a report to the FDA in 1969. Our interest here is largely in drugs that may be purchased without a physician's prescription, so-called over-the-counter (OTC), or nonprescription, drugs. The study covered 420 of the estimated 350,000 such products and declared that only one-quarter of the 420 examined were effective. These findings pointed out the need for a more comprehensive review of the efficacy of all OTC drugs, applying identical standards to every one. Obviously, it would be impossible to study all of the estimated one-third million products, so it was decided to examine only their 200 active ingredients.

A further obstacle needed to be overcome. Some of the older drugs had been "grandfathered" under both the 1938 act and the 1962 amendments since they were covered by the original 1906 Food and Drugs Act. How could the FDA remove these drugs from the market? Even if they were proven ineffective, they were apparently immune from the "effective" requirement established at the later dates.

The FDA reached these grandfathered drugs by what can only be described as an extremely innovative application of administrative law.[2] The agency simply declared that a drug would be considered misbranded if the manufacturer made any claims for it which were not in accord with the findings of one of seventeen panels set up in 1972 to review the efficacy of the active ingredients of all OTC drugs. In other words, a particular drug, even though exempt from proofs of safety and efficacy under existing laws, was barred from commerce if any unsubstantiated claim was made as part of the labeling that it was "good" for anything, that is, effective for the treatment of a disease state. The word *label* is very broadly interpreted to include not only the words printed on the container or package but also any literature accompanying it, such as a package insert.

In their search for proof of efficacy of the various OTC drugs, the seventeen panels, each of which concerned itself with a different class of therapeutic agents, had a number of potential sources of information. Neither testimonials nor market success were considered reliable criteria. This left in vitro tests (tests conducted outside the patient or any living organism) and various kinds of clinical trials (on patients) as the most acceptable methods.

Relatively simple in vitro tests may be suitable for the evaluation of a small number of certain drugs. For example, antacids can be mixed with acid to determine their neutralizing capacity. However, to prove the safety and efficacy of most therapeutic agents, pharmacological studies are necessary. They are ordinarily begun in small animals and continued with increasing complexity all the way through a series of so-called randomized, double-blind clinical trials in human beings. That simply means that some patients selected at random receive the drug while others are given a placebo; both groups are unaware of what they are taking. The results are evaluated by a physician who is also unaware of who is getting what treatment, and after statistical analysis of the data, a judgment is made of the drug's effectiveness.

Such studies, particularly the complex clinical trials, are extremely expensive. Estimates vary, but a 1990 study placed the total average cost of developing a single new chemical compound for drug use at $231 million over a period of twelve years![3] The most recent estimate for the average cost of a successful new drug application is $359 million.[4] A large part of this amount would certainly be the cost of testing the compound to establish its safety and efficacy as required in the New Drug Application.

Obviously, no company is going to make this kind of investment unless there is a reasonable expectation that it will be able to recover its costs and also show a profit. Since patent protection, now limited to a maximum of seventeen years, begins in the early stages of development of a drug, not when it is approved for marketing by the FDA, about 90 percent of the new drugs currently introduced have a remaining patent life of less than eleven years. The old "plant" drugs, some of which have been known and used for centuries, do not even qualify for this degree of protection. For this reason, the pharmaceutical industry has shown little interest in sponsoring studies on them, and the safety and efficacy of most herbal remedies remains unproven. Remember, too, that the panels did not necessarily review a drug unless requested to do so by a manufacturer or some other interested party who was then asked to provide quantities of supporting data.

In 1990, results of the FDA study of over-the-counter drugs, including many herbs, were released to the public. The results were very disappointing. Although some plant drugs such as cascara bark and senna leaves and pods were found to be safe and effective (in these cases as laxatives) and placed in category I, some 258 herbs and herbal products were not so judged. Of these, 142 were categorized as unsafe or ineffective, category II. There was insufficient evidence to evaluate the effectiveness of the other 116, and these were placed in category III.

Keep in mind that under this system, the FDA made judgments only on the basis of the evidence presented. Therefore, when it placed peppermint oil in category III, the FDA was not saying that peppermint oil was an ineffective digestive aid; it was merely noting that it had received insufficient evidence to allow peppermint oil to be designated effective. The same is true for prune concentrate and powder, which were not classified as effective laxatives even though they are universally known to be such.

The FDA has for some time maintained a list of substances "Generally Recognized as Safe," better known in the trade as the

GRAS list. About 250 herbs appear on this list, primarily based on their use as food additives, that is, as flavors or spices in the culinary arts and the beverage industry. Some of them, such as ginger and licorice, are also employed for their medicinal action, but GRAS listing does not mean that the FDA has approved the herb for therapeutic purposes.

All of these circumstances have led to the situation I mentioned before in which, until recently, practically all herbal remedies have been removed from the shelves of pharmacies and from the supervision of knowledgeable pharmacists. Such remedies migrated to the health food stores where they are sold under the guise of herbs, teas, health foods, food supplements, nutritional products, etc., labeled only with the name of the product. No claim of effectiveness for any condition appears on the label of such containers nor in any leaflet or advertisement that directly accompanies the drug. Any such claim would cause the product to be declared misbranded and render it subject to confiscation. How then does the uninformed consumer learn the uses of the various herbs? Salespeople generally avoid recommendations (especially if the customer looks like a law enforcement official), since such activity could result in a charge of unlicensed practice of medicine. However, the clerks in health food stores or the catalogs of mail-order establishments will refer interested persons to a large selection of books, pamphlets, and charts that list the drugs and describe their supposed uses. Some of these information sources are quite broad in scope and are called herbals or natural-medicine books. Others limit their coverage to a single drug or therapeutic class of drugs. Still others list the various diseases or conditions that require treatment and then recommend specific remedies. In this way, current laws and regulations requiring that OTC drugs be proven safe and effective prior to marketing are circumvented.

Labeling a package only with the common name of a drug has some very serious drawbacks aside from the omitted information on utility. The popular names of plants are not only numerous but inexact. Used without a qualifying adjective, the term "snakeroot" applies to at least six different plants, including *Actea pachypoda* Elliott, *Aristolochia serpentaria* L., *Asarum canadense* L., *Cimicifuga racemosa* (L.) Nutt., *Eupatorium rugosum* Houtt., and *Senecio aureus* L. Several modifying adjectives are also used to denote these or other species. Thus we have black snakeroot, button snakeroot, Canada snakeroot, corn snakeroot, heart snakeroot, Indian snakeroot, large snakeroot, prairie snakeroot, rattle snakeroot, Sampson's snakeroot, seneca (senega or seneka) snakeroot, Texas snakeroot, Virginia snakeroot, and white snakeroot, among others. When I see any plant material labeled snakeroot, I feel like a rider for the Pony Express watching the last telegraph pole being placed in position on the California line. It is depressing!

We see packages labeled oriental ginseng, wild red American ginseng, Korean ginseng, Tienchi ginseng, Chinese ginseng, and ginseng ad infinitum and ad nauseam, often with no indication of the botanical origin of the plant material inside. We see mistletoe on an herb tea label and wonder whether it is from the European species (*Viscum album* L.) or the American. The latter term could refer to as many as 200 different species of the genus *Phoradendron.* When it comes to gotu kola (which does not contain any cola), and Fo-ti-tieng, which is not the same as fo-ti, the situation becomes dangerously confusing.

The only answer to this herbal Tower of Babel is to require that the scientific name, that is, the Latin binomial, of the plant appear on the labels of all herbs. As a matter of fact, the American Society of Pharmacognosy has recommended to the FDA that in addition to the scientific name, the label should indicate the part(s) of the plant represented, the country of origin, and a specific lot number that could be related to a voucher specimen maintained for reference purposes. That would solve the identity problem; questions concerning safety, efficacy, and potency would still remain, of course.

One unfortunate part of this whole situation is that at least some of the drugs of natural origin which must legally be sold in this manner without any indication of use are no doubt safe and do possess useful therapeutic properties. We know this from the continued widespread use of these products in countries other than the United States. The science of herbal medicine is far more advanced in Germany today than in the United States, and the laws and regulations of that country are more realistic than ours with respect to the evidence required to prove safety and efficacy of an herbal product. There a doctrine of "reasonable certainty" prevails. This means that the health authorities place considerable emphasis on reports of the experiences of general practitioners in evaluating a plant drug, and such experiences are supplemented by evidence found in the literature and supplied by manufacturers.

For several years, a special committee (Commission E) of the *Bundesgesundheitsamt* (Federal Department of Health) reviewed the safety and efficacy of herbs and published the results of its findings in monographic form in the *Bundesanzeiger,* a publication equivalent to our *Federal Register.* These monographs constitute some of the most up-to-date and useful information extant on the safety and efficacy of plant drugs. They are frequently cited as the findings of German health authorities in the discussions of the various herbs that constitute the remainder of this volume.

To clarify the impact of the different regulatory philosophies in Germany and the United States, let us take a brief look at a single herb, valerian. The dried rhizome and roots (underground parts) of *Valeriana officinalis* L. have been valued as a tranquilizing agent for more than 1,000 years. Valerian enjoyed official status in the United States for 150 years and was included in *The United States Pharmacopeia* (USP) from 1820 to 1942 and in *The National Formulary* from 1942 to 1950. It is still available in pharmacies in the form of a tincture, but a pharmaceutical manufacturer would probably drop this product rather than spend the time and money necessary to prove its safety and efficacy—after ten centuries of use. However, valerian "dietary supplement" products—teas, capsules, tablets, extracts, and the like—abound in health food stores.

In contrast to the limited availability of pharmaceutical quality valerian in this country, more than 100 different proprietary drug products containing it or its active principles are currently marketed in Germany. Research carried out there has not yet been able to establish with certainty the exact identity of the active principles responsible for its calmative and sedative properties, but German health authorities have found the drug effective. They have also noted that it is free from side effects and contraindications and have established a proper schedule for its various dosage forms. Such a gentle tranquilizer as valerian has a distinct advantage over its synthetic cousins in that its effects are not synergistic with (are not usually potentiated by) alcohol. It is indeed unfortunate that this useful drug is primarily available in the United States as a foul-smelling tincture, bitter herbal tea, or capsule instead of a more palatable legitimate drug form. And this is entirely a result of present laws and regulations governing the sale of drugs in this country.

Fortunately, Mark Blumenthal, Executive Director of the American Botanical Council in Austin, Texas, has produced an English translation of the complete Commission E monographs. For the first time, a wide American audience will have access to the regulations produced by the German health authorities.

So much for the past and present. What about the future? Fortunately, there are some hopeful signs that this extremely negative legal and regulatory situation in the field of herbal medicine will not continue indefinitely. The groundswell of consumer enthusiasm for all things natural, including herbs, sometimes referred to as the "green wave," may develop enough influence to induce needed changes.[5]

An encouraging development has taken place in the herbal regulatory field in Canada. In 1984, the Health Protection Branch of that country convened an Expert Advisory committee to study the labeling of herbal products. The committee report, issued in 1986, recommended the establishment of a new class of remedies to be designated "Folklore Medicines."[6] This would include herbal products demonstrated to be safe, but whose efficacy was not necessarily proven by standard methods applicable to other drugs.

Although this particular recommendation was not implemented, the report set the stage for truly significant developments. The Health Protection Branch established regulatory procedures that permit health claims to be made for herbs on the basis of information available in standard reference sources (pharmacopeias, pharmacology books, etc.) or supportive references in the general scientific and clinical literature. The system allows the labeling of traditional medicines to inform the consumer regarding the therapeutic use of the product and the proper dosage.[7] The regulations are eminently sensible ones that might well be emulated in the United States. However, some believe that product liability laws and the litigious climate prevailing in this country might militate against enactment of such a plan, despite its many attractive features.

However, Canada, once considered to be on the international cutting edge of natural products regulation, has become another "average" regulatory environment for herbs with the closing of the Natural Products Section of the Bureau of Drug Research, Health, and Welfare in Canada in 1993. Recent decisions, such as that to ban any product containing gotu kola (*Centella asiatica*) have been described by a former regulatory official as "bewildering."[8]

Still, climates change and the wishes of a large group of enthusiasts who seek access to quality herbal products and accurate information about them cannot be disregarded indefinitely. Many are hopeful that, somewhere in the process, sensible regulations will be developed for herbs as well.

A step toward change has occurred since *The Honest Herbal,* Third Edition, was published. In 1994, a landmark legislative anomaly, The Dietary Supplement Health and Education Act of 1994 (DSHEA), passed by a voice vote in the waning hours of the 103rd Congress at 3 a.m. on Friday, October 7, 1994, just before final adjournment. The act passed by the Senate at 12:27 a.m. on Saturday, October 8, 1994, then was signed into law by President Clinton shortly therafter. As a result, most herb products are now sold in a food category, designated "dietary supplements."

Some legislative intents of DSHEA are to guarantee availability of products; allow truthful, nonmisleading scientific information to be used in conjunction with their sale; and give consumers some information on the product's benefits, as well as appropriate cautions. The act also places the burden of proof that a dietary supplement is adulterated or unsafe on the government, which must now present its evidence in federal court, rather than simply ordering a manufacturer to stop selling a product.

DSHEA in theory permits "third-party literature" such as publications, articles, chapters in books, and scientific reports to support the sale of dietary supplements, without being considered "drug" labeling. The information must not be false or misleading, nor may it promote a particular manufacturer or product brand; it must present a balanced view of the scientific information and, if displayed in a store, must be physically separate from the product and free of any appendages such as stickers. What exactly constitutes third-partly literature has yet to be defined in a regulatory context.

The act allows product labels to describe effects on general well-being or on "structure or function in humans," but drug claims may not be made. In mid-1998, proposed rules for "structure function" claims were published by the FDA in the *Federal Register*, which, by broadening the definition of disease, intend to limit severely what a manufacturer can say on a product label. One intent of DSHEA is to provide consumers with truthful, nonmisleading information on the use of herbal "dietary supplements." The proposed regulations, once again, seemingly attempt to restrict consumer access to information and are being challenged by industry.

Dietary supplement labels with structure or function claims must also carry a disclaimer: *This statement has not been evaluated by the Food and Drug Administration. This product is not intended to diagnose, treat, cure, or prevent any disease.* At least the first sentence in the required disclaimer is truthful.

DSHEA also established an Office of Dietary Supplements within the National Institutes of Health to conduct, coordinate, and collect data on dietary supplements and to advise the Secretary of Health and Human Services. A separate Presidential Commission on Dietary Supplement Labels was also formed through DSHEA to study and make recommendations on dietary supplement labels. The Report of the Commission on Dietary Supplement Labels was released on November 24, 1997. The seven-member commission's report strongly recommended that the FDA promptly establish a panel of experts to review nonprescription claims for botanical products that, for all practical purposes, are used as drugs. They recommended that regulatory systems of other countries should be reviewed for guidance in developing appropriate labeling and regulatory mechanisms. Both of these suggestions have been before the FDA for years, but without action.

DSHEA and its regulations seek to define herbs in the context of the modern marketplace, but it is only an interim legislative step. Interested consumers have already succeeded in convincing their elected representatives that they are not satisfied with a "we don't know" answer to the questions concerning whether catnip really has a useful sedative effect or if garlic actually reduces high blood pressure or if ginseng definitely increases resistance to disease. The only way to stimulate the necessary research required to answer such questions definitely is to make it financially profitable to obtain the answers. Easing the unnecessarily rigid standards for marketing plant drugs long in use as traditional remedies is, in my opinion, the best way to accomplish this objective. DSHEA is not the legislative act that will accomplish this goal. However, DSHEA did indirectly move herb products from the health food market into the mass market. By the year 2000, several large multinational pharmaceutical companies will undoubtedly offer herb products under the regulatory provisions of DSHEA. Market forces, driven by consumer demand, may create a competitive environment in which companies have to spend research dollars to survive. Time will tell. Personally, I look toward the herbal future with interest, enthusiasm, and modest optimism.

REFERENCES

1. D. R. Harlow. *Food, Drug, Cosmetic Law Journal* 32: 248-272, 1977.
2. M. C. Gerald. *American Pharmacy* NS19(5): 18-22, 1979.
3. Anon. *American Pharmacy* NS30(7): 10, 1990.
4. Anon. *Chemical Marketing Reporter,* August 9, 1993, p. 9.
5. V. E. Tyler. *Economic Botany* 40: 279-288, 1986.
6. Report of the Expert Advisory Committee on Herbs and Botanical Preparations. Minister of National Health and Welfare, Canada, 1986, 17 pp.
7. M. Blumenthal. *HerbalGram* 22: 18, 35, 1990.
8. D. V. C. Awang. Safety and Toxicity Issues Concerning Herbal Medicinal Uses: Mistaken Identity, Misinformation and Misregulation. Presentation for Botanical Medicine in Modern Clinical Practice, Columbia University, New York, May 26-29, 1988.

HERBS AND RELATED REMEDIES

Alfalfa

Anything green that grew out of the mould
Was an excellent herb to our fathers of old.

Rudyard Kipling
"Our Fathers of Old"
Stanza 1

Apparently our fathers of more recent vintage were the ones to ascribe therapeutic value to this plant. It is still difficult to understand how alfalfa, or lucerne, as it is known in Britain, ever gained a reputation as a medicinal herb. The leaves and flowering tops of *Medicago sativa* L. are not discussed in any of the classic American scientific works on natural drugs.[1,2] Even Grieve's *A Modern Herbal,* a comprehensive work of English origin, scarcely mentions the plant.[3] Of course, this perennial member of the family Fabaceae is one of our most common cultivated forage plants, being fed to animals either as hay or in a dehydrated form. While there is little or no research on alfalfa's health benefits for humans, there is a vast scientific literature on its value as a dietary supplement for lactating cows.

Travelers may be familiar with the odor and taste of alfalfa and not even know it. Anyone who has driven Interstate Highway 80 across Nebraska during the summer months and passed through Lexington and Cozad (the alfalfa capital of the world) has noted the peculiar green haze in the air and has smelled the pungent dust from the enormous alfalfa dehydrating plants that abound in that area.

Apparently someone in the recent past decided that if cattle could grow strong and healthy by eating alfalfa, the plant must have therapeutic value for human beings. That person was Alexander L. Blackwood, MD, of Chicago, a homeopathic physician, who evidently thought if it would fatten cows, it might fatten humans too. In the early twentieth century, it was introduced by both homeopathic and eclectic physicians as a remedy to "increase the appetite and the flesh." Its virtues were extolled by Ben A. Bradley, MD, as quoted in a booklet by Dr. Blackwood:

> I find in Alfalfa, after about seven years' clinical test in my practice and on myself, *a superlative restorative tonic*, but it does not act as a stimulant, after the manner of alcohol, cocaine or other habit-forming drugs. It rejuvenates the whole system by increasing the strength, vim, vigor, and vitality of the patient.[4]

Consequently, the herb is now widely advocated for consumption in the form of a tea or as tablets or capsules of the dried plant itself for a variety of ailments. We read testimonials to the efficacy of alfalfa tea in the cure of various types of arthritic conditions, including rheumatoid arthritis. Advocates also tell us that large quantities of alfalfa tablets taken before meals will prevent the absorption of cholesterol, thus benefiting our arterial blood flow and especially our heart.[5] Claims are made for the effectiveness of the tea in treating diabetes[6] and in stimulating the appetite and acting as a general tonic. A recent study evaluated the effect of an alfalfa aqueous extract (1mg/ml) on insulin release in streptozotocin-diabetic mice. At a dose of 62.5g/kg of body weight, an insulin-releasing and antihyperglycemic activity was observed.[7]

Scientific or clinical evidence in support of all but antidiabetic claims is scanty or totally lacking. There is one report that saponins of alfalfa root, which is not the part of the plant generally used, prevented an expected increase in plasma cholesterol in monkeys. Counterbalancing this, however, is evidence that alfalfa saponins are hemolytic and may interfere with the utilization of vitamin E.[8]

Because of its importance as an animal feed, alfalfa has been the subject of numerous and detailed chemical analyses. They have revealed the presence, in addition to the aforementioned saponins, of such constituents as fiber, protein, fats, minerals (calcium, phosphorus, iron, etc.), organic acids, vitamin K_1, a small amount of vitamin C, various pigments including chlorophyll, and the like.[9] Although some of these compounds do possess minor physiological activities, none is of significant therapeutic value, at least in the amount present in reasonable quantities of the herb. Considering the absolute lack of any proof of their utility in human medicine, alfalfa tablets, which presently may be purchased for a penny a piece, are still a bad buy. If you enjoy the taste of alfalfa sprouts in salads, they are refreshing and generally harmless, so feel free to eat them, at least in moderation.

There is good reason to insert the words "in moderation" in the last sentence. Since 1981, it has been recognized that eating very large quantities of alfalfa seeds daily could produce irreversible blood abnormalities (pancytopenia) in human beings.[10] Subsequent studies have shown that systemic lupus erythematosus (SLE), an inflammatory connective tissue disease, can be induced in normal monkeys by feeding alfalfa seeds or sprouts.[11] Also, persons suffering from clinically inactive SLE may have that condition reactivated by taking quantities of alfalfa tablets.[12] It seems likely that a nonprotein amino acid, L-canavanine, contained in alfalfa may play a role in causing the blood abnormalities and in inducing or reactivating SLE in persons having a predisposition to that condition. These latter individuals should be very cautious about consuming any alfalfa product, and since predisposition may not always be recognized, moderation seems generally advisable.

REFERENCES

1. H. W. Felter and J. U. Lloyd. *King's American Dispensatory,* Eighteenth Edition, Volumes 1-2. The Ohio Valley Company, Cincinnati, Ohio, 1898 and 1900.

2. H. W. Youngken. *Textbook of Pharmacognosy,* Sixth Edition. The Blakiston Company, Philadelphia, Pennsylvania, 1943.

3. M. Grieve. *A Modern Herbal,* Volume 2. Dover Publications, New York, 1971, pp. 501-502.

4. A. L. Blackwood. *Observations with Medicago sativa.* Lloyd Brothers, Cincinnati, Ohio, 1915, p. 4.

5. M. Bricklin. *The Practical Encyclopedia of Natural Healing.* Rodale Press, Inc., Emmaus, Pennsylvania, 1976, pp. 28-29, 202-203.

6. R. Adams and F. Murray. *Health Foods.* Larchmont Books, New York, 1975, pp. 145-147.

7. A. M. Gary and P. R. Flatt. *British Journal of Nutrition* 78(2): 325-334, 1997.

8. A. Y. Leung and S. Foster. *Encyclopedia of Common Natural Ingredients Used in Food, Drugs, and Cosmetics,* Second Edition. John Wiley and Sons, New York, 1996, pp. 13-16.

9. P. H. List and L. Hörhammer (Eds.). *Hagers Handbuch der Pharmazeutischen Praxis,* Fourth Edition, Volume 5. Springer-Verlag, Berlin, 1976, pp. 732-734.

10. M. R. Malinow, E. J. Bardana Jr., and S. H. Goodnight Jr. *Lancet* I: 615, 1981.

11. M. R. Malinow, E. J. Bardana Jr., B. Pirofsky, S. Craig, and P. McLaughlin. *Science* 216: 415-417, 1982.

12. J. L. Roberts and J. A. Hayashi. *New England Journal of Medicine* 308: 1361, 1983.

Aloe

. . . the public must learn how to cherish the nobler and rarer plants, and to plant the aloe . . .

Margaret Fuller

Although she was thinking of the plant in another connection when she wrote these lines, Margaret Fuller's advice about aloe may be taken quite literally by those seeking a handy, homegrown remedy for minor burns, abrasions, and other skin irritations. There is a vast folk literature indicating that the fresh gel or mucilage of *Aloe barbadensis* Mill. (family Liliaceae), otherwise known as *A. vera* (L.) Webb & Berth. or *A. vulgaris* Lam., promotes wound healing on external application. It has also been taken internally for a variety of maladies.[1,2]

A continuing source of confusion to persons interested in herbs is the fact that aloe is the source of two products that are completely different in their chemical composition and their therapeutic properties but which have very similar names that are sometimes interchanged. Aloe (aloe vera) gel or mucilage is a thin, clear, jellylike material obtained from the so-called parenchymal tissue making up the inner portion of aloe leaves. It is prepared from the leaf by various procedures, all of which involve its separation not only from the inner cellular debris but, especially, from specialized cells known as pericyclic tubules that occur just beneath the epidermis or rind of these same leaves.[3] Such cells contain a bitter yellow latex or juice that is dried to produce the pharmaceutical product known as aloe, an active cathartic.

Aloe gel (mucilage) is used both externally and internally for its wound-healing properties and as a general tonic or cure-all. This is the aloe product commonly incorporated in a wide variety of non-laxative drug and cosmetic products. Aloe latex or juice, usually in its dried form, is employed as a potent cathartic. Unfortunately, the mechanical separation processes employed are often not completely effective. As such, aloe gel is sometimes contaminated with aloe latex, thus inducing an unwanted laxative effect following consumption of the so-called gel. In addition, advertisements prepared by copywriters who do not understand the vast difference between aloe gel and aloe juice often use the word juice to describe the thin mucilaginous gel.

To confuse matters even more thoroughly, there is still another product called aloe that is entirely different from the two just described. That is the aloe of the Bible, the so-called lignaloes or aloe wood, a fragrant wood from an entirely different plant that was once used as an incense.[4] It has nothing to do with the aloe we are discussing except that some persons try to glamorize aloe gel by incorrectly ascribing to it a biblical origin. The names may be the same, but the plants referred to are not. Actually, aloe latex has been used as a laxative for about eighteen centuries, but neither it nor aloe gel is referred to in the Bible.

Having disposed of these nomenclatural difficulties, let us return to the use of aloe gel (mucilage) as a wound-healing agent and all-around remedy. Although many sources agree that the gel possesses some activity in its fresh state, there is controversy over whether this activity is retained during storage. Commercial processors claim that the stability problem has been overcome, and a "stabilized" product is incorporated in a wide variety of preparations, including juices, gels, ointments, creams, lotions, and shampoos.[5]

However, at least one scientific test failed to verify any beneficial effects of a "stabilized" aloe vera gel on human cells.[6] Fluid from fresh leaf sources was found to promote significantly the attachment and growth of normal human cells grown in artificial culture. It also enhanced the healing of wounded monolayers of the cells. On the other hand, the "stabilized" commercial product not only failed to induce such effects but actually proved toxic to the cultured cells. The investigators who carried out these studies concluded that commercially prepared aloe vera gel fractions "can markedly disrupt the *in vitro* attachment and growth of human cells."

Review of several other studies led to the conclusion that a number of them did provide evidence to support the use of aloe vera gel, and some preparations containing it, for the treatment of various types of skin ulceration in humans and burn and frostbite injuries in animals.[7] More recently, a cream base containing aloe was found effective in preserving circulation in the skin after frostbite injury.[8] Stabilized aloe vera was shown to produce a dramatic acceleration of wound healing in patients who had undergone full-face dermabrasion.[9]

It is postulated that aloe may function in such cases by inhibiting bradykinin, a pain-producing agent; also, it apparently hinders the formation of thromboxane, whose activity is detrimental to burn wound healing. Aloe gel also has antibacterial and antifungal properties. Studies on the mechanism of action of aloe gel or partially purified extracts in vitro on skin wound-healing repair processes provides evidence that aloe stimulates fibroblast and epithelial cell growth, induces lectinlike responses in human immune cells, and stimulates neuronlike cell growth.[10] Still, relatively little is known about the identity and stability of the ingredients responsible for these effects. A glycoprotein fraction has been shown to promote cell growth in human and animal cell media, while a polysaccharide fraction did not stimulate growth.[11] Many compounds of aloe are probably subject to deterioration on storage, so use of the fresh gel is the only way to be certain of maximal activity.

In addition, various commercial preparations often contain minimal amounts of aloe. One way to determine the relative quantity present is to determine the position of aloe in the list of ingredients stated on the label. If it is not near the top, the amount present is probably quite small. Also, be cautious about preparations labeled "aloe vera extract," which may be highly diluted or "reconstituted aloe vera," meaning that the product has been prepared from a powder or liquid concentrate.[12]

Aloe gel (often incorrectly designated "juice") is described in the popular literature as a cleanser, anesthetic, antiseptic, antipyretic, antipruritic, nutrient, moisturizer, and vasodilator and is also said to possess anti-inflammatory properties and to promote cell proliferation. Recommendations for internal use range from the treatment of coughs to constipation; externally it is used primarily for burns, for conditioning the skin, and even for headache. A salesman drinks it to "detoxify" his system. One Arkansas physician applied it to relieve the symptoms of poison ivy. The utility of aloe in treating many of these conditions has not been verified.[13]

Mixed results have been published on the traditional use of aloe juice in treating diabetes. A controlled clinical study involved seventy-seven volunteers who were administered one tablespoonful of aloe juice, or placebo, twice a day for up to forty-two days. A significant reduction in blood sugar and triglyceride levels was observed in the treatment group. Cholesterol levels remained the same in both groups.[14] This conflicts with an earlier study that was unable to support benefit of claimed efficacy in diabetes mellitus or in gastric ulcers.[15]

In the 1990s, a body of scientific literature has arisen providing a rational scientific basis for aloe's use in treating minor wounds and burns. This provides a foundation to support an impressive body of folklore attesting to aloe's healing properties on external application.[16]

Many people keep a potted aloe plant on the windowsill in the kitchen so that a leaf can be cut off and the freshly exuded gel applied to minor burns. Since the safety of such procedures has never been questioned, it is a therapy that has much to recommend it. Also, the treatment is inexpensive and overcomes the potential problems of stability and retention of the gel's desirable properties following commercial processing and storage.

REFERENCES

1. J. F. Morton. *Economic Botany* 15: 311-319, 1961.

2. R. H. Cheney. *Quarterly Journal of Crude Drug Research* 10: 1523-1530, 1970.

3. D. L. Smothers. *Drug and Cosmetic Industry* 132(1): 40, 77-80, 1983.

4. J. U. Lloyd. *Origin and History of All the Pharmacopeial Vegetable Drugs, Chemicals and Preparations,* Volume 1. The Caxton Press, Cincinnati, Ohio, 1921, pp. 4-14.

5. J. Flagg. *American Perfumer and Aromatics* 74(4): 27-28, 61, 1959.

6. W. D. Winters, R. Benavides, and W. J. Clouse. *Economic Botany* 35: 89-95, 1981.

7. A. D. Klein and N. S. Penneys. *Journal of the American Academy of Dermatology* 18: 714-720, 1988.

8. R. L. McCauley, J. P. Heggers, and M. C. Robson. *Postgraduate Medicine* 88(8): 67-68, 73-77, 1990.

9. J. E. Fulton, Jr. *Journal of Dermatologic Surgery and Oncology* 16: 460-467, 1990.

10. C. F. Bouthet, V. R. Schirf, and W. D. Winters. *Psychotherapy Research* 9: 185-188, 1995.

11. A. Yagi et al. *Planta Medica* 63: 18-21, 1997.

12. T. R. Fox. *Health Foods Business* 36(12): 45-46, 1990.

13. *Lawrence Review of Natural Products,* April 1992.

14. S. Yongchaiyudha, V. Rungpitarangsi, N. Bunyapraphatsara, and O. Chokechchaijaroenporn. *Phytomedicine* 3(3): 241-243, 1997.

15. M. W. L. Koo. *Psychotherapy Research* 8: 461-464, 1994.

16. V. E. Tyler. *Hoosier Home Remedies.* Purdue University Press, West Lafayette, Indiana, 1985, p. 30.

Angelica

All parts of the tall, perennial, herbaceous plant *Angelica archangelica* L. (family Apiaceae) contain a very pleasant-smelling aromatic volatile oil, which probably accounts for the continued use of the root, fruits, and leaves of this and other closely related species of *Angelica* in folk medicine.[1] *Angelica atropurpurea* L. is the one commonly employed in the United States.

The drug has been recommended as an antiflatulent (antigas treatment), a diuretic, a diaphoretic (sweat producer), and a counterirritant. It has also acquired some reputation as an emmenagogue (promotes menstrual flow) and abortifacient. A recent study on the effect of the essential oil on skeletal and smooth muscle contractility in vitro failed to show any response.[2] There is no proof that the drug is particularly effective in any of these applications; severe poisoning has resulted from large doses of the root administered in an attempt to induce abortions.

At present, the most important application of angelica root and seed is in the flavoring of various alcoholic beverages. It is a component of a number of herb liqueurs, such as Benedictine and Chartreuse. Together with juniper berries and coriander seed, angelica root is one of the principal flavoring ingredients in gin.[3]

The purplish stems of angelica are sometimes collected and "crystallized" with sugar to make a pleasant-tasting confection. Those who consume them or any other parts of angelica should be aware that the plant contains, in addition to the fragrant volatile oil, a number of furocoumarins, e.g., angelicin, bergapten, imperatorin, and xanthotoxin. On contact with the skin, these so-called psoralens may induce photosensitivity (to the sun), resulting in a kind of dermatitis.[4] Studies have shown that these compounds are photocarcinogenic (cancer-causing) in laboratory animals and are acutely toxic and mutagenic even in the absence of light. Investigators have now concluded that psoralens present risks of such magnitude to humans that unnecessary exposure to them (via consumption or contact) should be avoided.[5]

Photochemotherapy of psoriasis with psoralens and long-wave ultraviolet light is sometimes used with dramatic results. However, because of the potential toxicity of the psoralens, the treatment is reserved for severe, disabling psoriasis that does not respond to other therapies.[6] For the same reason, the use of angelica as a self-selected drug cannot be recommended. However, angelica root is approved in Germany for the treatment of loss of appetite, flatulence, plus a feeling of fullness and mild gastrointestinal tract spasms leading to peptic discomfort.[7] The basis of the Commission E's recommendations is unknown. To make matters even worse, novice collectors may also confuse angelica with water hemlock (*Cicuta maculata* L.), an extremely poisonous plant.[8]

REFERENCES

1. H. W. Youngken. *Textbook of Pharmacognosy,* Sixth Edition. The Blakiston Co., Philadelphia, Pennsylvania, 1948, pp. 623-625.

2. M. Lis-Balchin and S. Hart. *Journal of Ethnopharmacology* 58(3): 183-187, 1997.

3. E. Guenther. *The Essential Oils,* Volume 4. D. Van Nostrand Co., New York, 1950, pp. 553-563.

4. P. H. List and L. Hörhammer (Eds.). *Hagers Handbuch der Pharmazeutischen Praxis,* Fourth Edition, Volume 3. Springer-Verlag, Berlin, 1972, pp. 88-98.

5. G. W. Ivie, D. L. Holt, and M. C. Ivey. *Science* 213: 909-910, 1981.

6. K. F. Swingle and D. C. Kvam. In *Modern Pharmacology,* Third Edition. C. R. Craig and R. E. Stitzel (Eds.). Little, Brown and Company, Boston, Massachusetts, 1990, pp. 622-623.

7. M. Blumenthal, W. R. Busse, A. Goldberg, J. Gruenwald, T. Hall, C. W. Riggins, and R. S. Rister (Eds.). *German Commission E Monographs. Therapeutic Monographs on Medicinal Herbs.* American Botanical Council, Austin, Texas, 1998.

8. W. C. Muenscher. *Poisonous Plants of the United States.* The Macmillan Co., New York, 1951, p. 175.

Apricot Pits (Laetrile)

Fraudulent and ineffective cancer cures are as old as the disease itself, and the intervention of well-meaning but medically naive "politicians" to treat cancer stems from an early date as well. In 1748, the House of Burgesses of the Commonwealth of Virginia undertook a study of one Mary Johnson's herbal "receipt of curing cancer." As a result of the anecdotal testimony provided, the House voted Mrs. Johnson a reward of 100 pounds. Similarly, in 1964, fifty-six U.S. Congressmen cosponsored a resolution, which fortunately failed to pass, authorizing the expenditure of $250,000 for the study of another unproven cancer remedy, krebiozen.[1]

During the 1970s, another fake cancer cure, laetrile, began to be widely promoted and used. Despite the fact that absolutely no scientific evidence existed for its therapeutic efficacy, demand for the product became so great that by 1981 the legislatures of some twenty-three states had legalized its use.[2] This whole sordid history, highlighted by (but not restricted to) Mary Johnson's "receipt," krebiozen, and laetrile, apparently indicates that in the absence of a cure for a terminal disease, desperate people want hope, not facts.

Laterile, as originally patented in the United States in 1961, is not the same compound as the one called laetrile today.[3] The former was technically known as mandelonitrile glucuronide, but it was relatively difficult to procure. Consequently, a closely related compound, amygdalin (mandelonitrile β-gentiobioside), became the laetrile of commerce. It is also sometimes referred to as vitamin B_{17}, although it is definitely not a vitamin. Amygdalin occurs naturally in a number of plant materials; however, the usual commercial source is the kernel of various varieties of *Prunus armeniaca* L. (family Rosaceae), commonly referred to as apricot pits. These vary appreciably in their amygdalin or laetrile content, which may reach 8 percent, but the kernels of some wild varieties contain twenty times as much as those of cultivated varieties of apricots. Serious cases of poisoning, especially among children, have been reported as a result of eating quantities of these seeds.[4]

Advocates of laetrile therapy for cancer believe that an enzyme, β-glucosidase, capable of breaking down the laetrile to release toxic cyanide, exists in large amounts in tumorous tissue but only in small quantities in the rest of the body. They further hypothesize that another enzyme, rhodanese [sic], which has the ability to detoxify cyanide, is present in normal tissues but deficient in cancer cells. These two factors supposedly combine to effect a selective poisoning of cancer cells by the cyanide released from the laetrile, while normal cells and tissues remain undamaged.[5] Needless to say, no scientific proof of this so-called mechanism of action has ever been presented.

Lacking any proof of safety and efficacy, the Food and Drug Administration banned laetrile from interstate commerce in 1971. However, several state legislatures, reacting to political pressures, legalized intrastate sale and use of the product. In 1980, similar pressures forced the National Cancer Institute to begin a clinical study of laetrile in terminal cancer patients.[6]

Conducted in collaboration with four major U.S. medical centers, the clinical tests showed that laetrile failed on four counts: it did not make cancer regress; it did not extend the life span of cancer patients; it did not improve cancer patients' symptoms; and it did not help cancer patients to gain weight or otherwise become more physically active.[7] Laetrile and natural products containing it, such as apricot pits, were thus found to be "ineffective as a treatment for cancer."

In the decade following this report, apricot pits and laetrile have practically disappeared from the American herbal scene. Nevertheless, it is useful to review occasionally the history of such a fraudulent cancer cure to remind us that desperate people desiring to cure the incurable are especially liable to fall victim to unprincipled charlatans whose motivation is basically financial, not altruistic. These persons will continue to advocate such fraudulent drugs, at least until the next bigger, better, and newer quack cure comes along. It further serves to remind us that, with respect to medicines, popularity does not necessarily equate with efficacy, nor legislative endorsement with scientific fact. There is nothing new to add.

REFERENCES

1. R. N. Grant and I. Bartlett. In *Unproven Methods of Cancer Management.* American Cancer Society, New York, 1971, pp. 1-2.
2. *The Indianapolis Star,* February 13, 1981, p. 23.
3. C. Fenselau, S. Pallante, R. P. Batzinger, W. R. Benson, R. P. Barron, E. B. Sheinin, and M. Maienthal. *Science* 198: 625-627, 1977.
4. J. W. Sayre and S. Kaymakcalan. *New England Journal of Medicine* 270: 1113-1115, 1964.
5. D. L. Poulson. *Herbalist* 4(4): 2-5, 1979.
6. T. H. Jukes. *Journal of the American Medical Association* 242: 719-720.
7. *Science News* 119: 293-294, 1981.

Arnica

The medicinal virtues of arnica were independently discovered by Europeans before the end of the sixteenth century and by American Indians at an early, but uncertain, date. Originally, the entire plant including the roots was employed, often internally, for a variety of conditions. Subsequently, the flower heads alone began to be used, either as a tincture (dilute alcoholic solution) or an ointment. The European drug is obtained from *Arnica montana* L.; the American product comes from *Arnica fulgens* Pursh, *A. sororia* Greene, *A. latifolia* Bong., and *A. cordifolia* Hook. All are closely related perennial herbs of the family Asteraceae with orange-yellow daisy-like flower heads. They are native to the meadows and mountainous regions of Europe and western North America.[1]

Writers on herbs recommend the application of arnica externally to reduce the inflammation and pain of bruises, aches, and sprains.[2] They often provide directions for the formulation of various arnica preparations intended for such external use.[3] Most discourage internal use of the drug, pointing out its involvement in severe, even fatal, cases of poisoning.

Scientific studies of the effects of alcoholic extracts of arnica on the heart and circulatory system of small animals have verified the folly of using the drug internally for self-medication.[4] Not only did it exhibit a toxic action on the heart, but in addition, it caused very large increases in blood pressure. Additional studies have confirmed arnica's cardiac toxicity.[5]

External application of the drug is quite a different matter. Although widely used as a home remedy for aches and bruises, no one could explain how or why arnica worked, if indeed it did. Sollmann, a respected pharmacologist, even speculated that the main active ingredient in arnica tincture was perhaps the alcohol.[6] Chemical studies isolated and identified large numbers of constituents,[7] but none of them accounted for the drug's reputed anti-inflammatory and analgesic effects.

Finally, in 1981, a report from Germany, where more than 100 different drug preparations containing arnica extract were then marketed, revealed that certain sesquiterpenoid lactones were the active principles.[8] Helenalin, dihydrohelenalin, as well as esters of these two compounds possess pharmacologic properties that explain a number of the actions of arnica. Besides producing anti-inflammatory and analgesic effects, the compounds also display some antibiotic activity. One drawback must be noted, however. Helenalin is an allergen and causes contact dermatitis in some persons. The potential for allergenic reactions is believed to be related both to the helenalin concentration and the delivery medium. Contact dermatitis occurs in individuals with sensitivity to helenalin-type sesquiterpenes. If this occurs, the application of arnica should immediately be discontinued. Arnica's risk-benefit ratio has recently come under question, although it is approved in Germany for external use in the treatment of hematomas, sprains, bruises, contusions, rheumatic pains of muscles and joints, and for fracture-related edema.[9]

Although more scientific studies are needed to define more precisely the physiologic properties of helenalin and related compounds in arnica, there is considerable rationale for its external application to reduce the inflammation and pain of various aches and bruises. The commercially available arnica creams or tinctures are best utilized for this purpose. Such use is vividly described in a turn-of-the-century poem about the damage inflicted on a candidate during an overly enthusiastic lodge initiation ceremony:

> The house is full of arnica,
> We do not dare to run about
> Or make the slightest sound
>
> We leave the big piano shut
> And do not strike a note;
> The doctor's been here seven times
> Since father rode the goat.
>
> He joined the Lodge a week ago—
> Got in at four A.M.,
> And sixteen brethren brought him home,
> Though he says he brought them.

His wrist was sprained and one big rip
Had rent his Sunday coat—
There must have been a lively time
When father rode the goat.

REFERENCES

1. A. Osol and G. E. Farrar Jr. *The Dispensatory of the United States of America,* Twenty-Fourth Edition. J. B. Lippincott, Philadelphia, Pennsylvania, 1947, pp. 98-100.
2. M. Grieve. *A Modern Herbal,* Volume 1. Dover Publications, New York, 1971, p. 55.
3. W. H. Hylton (Ed.). *The Rodale Herb Book.* Rodale Press Book Div., Emmaus, Pennsylvania, 1974, pp. 351-352.
4. A. W. Forst. *Naunyn-Schmiedebergs Archiv fur experimentelle Pathologie und Pharmakologie* 201: 242-260, 1943.
5. W. Werner. *Deutsche Apotheker Zeitung* 121: 199, 1981.
6. T. Sollmann. *A Manual of Pharmacology,* Seventh Edition. W. B. Saunders, Philadelphia, Pennsylvania, 1948, p. 146.
7. A. Y. Leung and S. Foster. *Encyclopedia of Common Natural Ingredients Used in Food, Drugs, and Cosmetics,* Second Edition. John Wiley and Sons, New York, 1996, pp. 40-42.
8. W. Werner, op. cit., p. 121: 199, 1981.
9. V. Schulz, R. Hänsel, and V. E. Tyler. *Rational Phythotherapy. A Physicians' Guide to Herbal Medicine,* Third Edition. Springer, Berlin, 1998, pp. 260-262.

Barberry

One of the confusing curiosities of crude drug nomenclature is the fact that barberry or berberis is obtained from plants of the genus *Mahonia* and not from species of *Berberis.* The reason for this is that the several species which yield the rhizome and roots (underground parts) constituting this drug were once classified as *Berberis* species but are now placed in the genus *Mahonia.* They include *M. aquifolium* (Pursh) Nutt. and *M. nervosa* (Pursh) Nutt., both commonly referred to as Oregon grape. These attractive members of the family Berberidaceae are evergreen shrubs with holly-like leaves and bluish black berries; *M. aquifolium* is generally taller (three feet plus) than *M. nervosa* (up to two feet). Common barberry, *Berberis vulgaris* L., was not a recognized source of the drug when it had official status, but the bark of its root and stem contains similar active principles and is also used similarly.[1] Claims for use of the fruits, bark, and root are not allowed in Germany, as the claimed effectiveness for various traditional uses is not scientifically documented.[2]

Barberry was reportedly used by the American Indians in cases of general debility and to improve the appetite. When the early settlers observed this, they employed the root as a bitter tonic. In addition, it was said to be of value as a treatment for ulcers, heartburn, and stomach problems when given in small doses; large doses have a cathartic effect.[3]

A number of isoquinoline alkaloids, especially berberine, berbamine, and oxyacanthine, account for the physiological activity of barberry. Several of the alkaloids exhibit antibacterial properties, and berberine is also effective against both amoeba and trypanosomes. This alkaloid has some anticonvulsant, sedative, and uterine-stimulant properties as well. Berbamine produces a hypotensive effect (lowers blood pressure).[4]

In light of the lack of anti-inflammatory response from the alkaloids in the bark, two phenolic compounds, 3-hydroxy-4-methoxyphenylethyl alcohol and syringaresinol, have been isolated from methanolic extracts, to which possible anti-inflammatory activity has been attributed.[5]

Despite these various properties, barberry and its contained alkaloids, which berberine is the principal one, are not very useful drugs. Including the crude plant material in bitter tonics has been essentially discontinued. Berberine salts continued to be used for some time in eyedrops because of their astringent properties, but this, too, has now ceased.

Never adopted as an official remedy in German phytomedicine barberry did come to the attention of a number of German physicians in the 1940s and 1950s and was reportedly used with positive results with a number of skin conditions, notably psoriasis vulgaris. A German research group recently conducted a randomized, placebo-controlled clinical trial to evaluate safety and efficacy of a topical *Mahonia aquifolium* bark extract in psoriasis. The study, conducted from the autumn of 1990 to the spring of 1992, involved eighty-two patients recruited by twenty-two physicians. Psoriasis vulgaris is a hereditary condition, which according to the authors, affects 1 to 3 percent of the European population. Patients were assessed after an average of four weeks of treatment. Subjective results were recorded by patients and physicians. Four patients reported adverse reactions to the ointment such as itching or burning, which were characterized as allergic reactions to the ointment. Based on this one study, the authors concluded that the berberis extract was "a potent and safe therapy of moderately severe cases of psoriasis vulgaris."[6]

More research is obviously needed. Any serious discussion of the drug, beyond potential topical applications, belongs more properly in the history books, not in medical books.

REFERENCES

1. A. Osol and G. E. Farrar Jr. (Eds.). *The Dispensatory of the United States of America,* Twenty-Fourth Edition. J. B. Lippincott, Philadelphia, Pennsylvania, 1947, pp. 1361-1363.

2. M. Blumenthal (Ed.). *German Commission E Monographs: Therapeutic Monographs on Medicinal Herbs.* American Botanical Council, Austin, Texas, 1998.

3. W. H. Hylton (Ed.). *The Rodale Herb Book.* Rodale Press Book Div., Emmaus, Pennsylvania, 1974, pp. 352-353.

4. A. Y. Leung and S. Foster. *Encyclopedia of Common Natural Ingredients Used in Food, Drugs, and Cosmetics,* Second Edition. John Wiley and Sons, New York, 1996, pp. 66-67.

5. C. Wirth and H. Wagner. *Phytomedicine* 4(4): 375-358, 1997.

6. M. Weisenauer and R. Lüdtke. *Phytomedicine* 3(3): 231-235, 1996.

Bayberry

The bayberry plant, *Myrica pensylvanica* Loisel, of the family Myricaceae, is a deciduous shrub widely distributed throughout the eastern and southern states. It is closely related to the wax myrtle, *Myrica cerifera* L., a somewhat larger evergreen shrub or tree that is also referred to as southern bayberry. Both species produce small bluish white berries, the wax of which is used to make the fragrant smelling bayberry candles, popular at Christmastime.

In folk medicine, the root bark of both species is administered internally, usually in the form of a warm infusion or tea, for its tonic, stimulant, and astringent properties. It is reputed to be especially valuable in the treatment of diarrhea. In large doses, bayberry bark acts as an emetic due to its irritating action on the stomach. The drug has also been used to increase the secretion of nasal mucus during head colds.[1] Applied in the form of poultices, the root bark is said to be useful in the treatment of chronic, so-called indolent ulcers. One herbalist describes the drug as, "If not absolutely the most useful article in botanic practice, it is certainly nearly so."[2]

Its continued use as a folk medicine directly evolves from its promotion by Samuel Thomson (1769-1843), a self-taught physician whose practice advocated use of emetics, stimulants, and purgatives. The course of treatment (the same for all diseases) included doses of lobelia, then cayenne pepper, followed by Thomson's No. 3 drug in the regime, bayberry, used "to scour the stomach and bowels and remove the canker." "It is an excellent medicine," Thomson extols, "either taken by itself or compounded with other articles; and is the best thing for canker of any article I have ever found. . . . When the stomach is very foul, it will frequently operate as an emetic."[3]

The nature of much "current" herbal information may be deduced from books published as recently as 1980, which continue to list only those constituents of bayberry originally determined in 1863 by analytical procedures now thought extremely primitive.[4] These compounds include an acrid and an astringent resin, tannic acid, gallic acid, and a principle called myricinic acid, which has never been characterized chemically.

Subsequent chemical investigations have identified several interesting chemical compounds in bayberry root bark. Three triterpenes, myricadiol, taraxerol, and taraxerone, are present in the drug plus a flavonoid glycoside, myricitrin. Of these compounds, myricadiol has been reported to have mineralocorticoid activity, that is, it influences sodium and potassium metabolism in the same way as the steroid principles of the adrenal cortex. Myricitrin has been shown to function as a choleretic (stimulates flow of bile) and as an agent toxic to bacteria, paramecia, and sperm.[5]

However, even if bayberry were a useful drug for any particular condition—a hypothesis that remains unproven—its safety, at least in large doses, is still in doubt because of the potential carcinogenic nature of its contained tannin. Injection of bark extracts into rats produced a significant number of malignant tumors during a relatively long-term (seventy-eight weeks) experiment.[6] These results raise a question about the safety of bayberry for consumption by human beings. Since the root bark has no proven medicinal value anyway, it seems best to restrict the use of the plant to its berries, whose wax does make nice-smelling candles.

REFERENCES

1. G. M. Hambright. *American Journal of Pharmacy* 35: 193-202, 1863.

2. R. C. Wren and R. W. Wren. *Potter's New Cyclopaedia of Botanical Drugs and Preparations,* New Edition. Health Science Press, Hengiscote, England, 1975, p. 30.

3. S. Thomson. *New Guide to Health; or, Botanic Family Physician.* J. Q. Adams, Boston, 1835, p. 55.

4. D. G. Spoerke Jr. *Herbal Medications.* Woodbridge Press Publishing Co., Santa Barbara, California, 1980, pp. 29-30.

5. B. D. Paul, G. Subba Rao, and G. J. Kapadia. *Journal of Pharmaceutical Sciences* 63: 958-959, 1974.

6. G. J. Kapada, B. D. Paul, E. B. Chung, B. Ghosh, and S. N. Pradhan. *Journal of the National Cancer Institute* 57: 207-209, 1976.

Betony

Betony or Wood Betony is one of those medicinal plants once believed to be good for practically everything, whose use in folk medicine decreased over the years until it is now thought to be of relatively little value. Its earlier importance is indicated by two old proverbs or sayings: Both the Italian, "Sell your coat and buy betony," and the Spanish, "He has as many virtues as betony," emphasize its former versatility as a remedy.

The drug consists of the dried herb, that is, the entire overground portion of the plant *Stachys officinalis* (L.) Trev., a square-stemmed perennial of the family Lamiaceae with a rosette of hairy leaves and a spike of pink or purplish flowers that attains a height of up to three feet. It is native to the cleared areas and meadows of Europe and is widely cultivated in herb gardens.[1]

In Roman times, the plant was thought to be a sure cure for forty-seven different diseases. (There is no need to list them because any one you can think of was probably included.) During the Middle Ages, many magical properties, including power against evil spirits, were attributed to betony.[2] Today the drug is still highly valued in folk medicine but principally for its properties as an astringent in treating diarrhea and irritations of the throat, mouth, and gums. An infusion or tea prepared from the leaves is either drunk or used as a gargle or mouthwash, according to the condition being treated.[3]

Betony contains about 15 percent tannin, which explains its use and its effectiveness as an astringent drug.[4] A Russian study found a mixture of glycosides in the plant, at least one of which was a flavonoid pigment. These glycosides were reported to have hypotensive (lower blood pressure) effects.[5] Although the report requires verification, it might partially explain the supposed effectiveness of betony in treating mild anxiety states and headache. A recent study found six new phenylethanoid glycosides, named betonyosides A through F, from the aerial parts of *Stachys officinalis*, although their contribution to biological activity is unknown.[6]

There are over 300 species of *Stachys*. Current interest in their use as folk medicines is primarily limited to Russia. Of the thirty-seven species indigenous to Russia, twelve are used as traditional medicines for their antiphlogistic, cholagogue, sedative, hypotensive, their wound-healing, and hemostatic properties.[7]

One species, *Stachys arvensis*, is associated with nervous and muscular locomotor disorders ("staggers") in sheep in Australia, characterized by limb paresis with knuckling of the fetlocks.[8]

Still, the only known utility of the drug is its astringent action, due to the tannins, which makes it effective in treating diarrhea and various irritations of the mucous membranes. In normal usage, betony should not cause any notable side effects, but overdosing may result in excessive irritation of the stomach.

REFERENCES

1. M. Stuart (Ed.). *The Encyclopedia of Herbs and Herbalism.* Grosset and Dunlap, New York, 1979, pp. 266-267.
2. M. Grieve. *A Modern Herbal,* Volume 1. Dover Publications, New York, 1971, pp. 97-99.
3. M. Pahlow. *Das Grosse Buch der Heilpflanzen.* Gräfe and Unzer GmbH, Munich, 1979, p. 83.
4. P. H. List and L. Hörhammer (Eds.). *Hagers Handbuch der Pharmazeutischen Praxix,* Fourth Edition, Volume 6B. Springer-Verlag, Berlin, 1979, pp. 506-507.
5. T. V. Zinchenko and I. M. Fefer. *Farmatsevtichnii Zhurnal* (Kiev) 17(3): 35-38, 1962.
6. T. Miyase, R. Yamamoto, and A. Ueno. *Phytochemistry* 43(2): 475-479, 1996.
7. V. G. Kartsev, N. N. Stepanichenko, and S. Auelbekov. *Chemistry of Natural Compounds* 30(6): 645-654; through *Review of Aromatic and Medicinal Plants* 2(3), abstract: 1210, 1996.
8. C. A. Bourke. *Australian Veterinary Journal* 72(6): 228-234, 1995.

Black Cohosh

Black cohosh consists of the underground parts (rhizome and roots) of the showy North American forest plant *Cimicifuga racemosa* (L.) Nutt., family Ranunculaceae. The plant's common names are numerous and include black snakeroot, rattleweed, rattleroot, bugbane, bugwort, and squaw root (not to be confused with blue cohosh). The genus *Cimicifuga* contains twenty-three temperate climate species: six from North America, one in Europe, the remainder from temperate eastern Asia. Similar to black cohosh, several Asian species are traditionally used for gynecological conditions.

The drug was introduced into medicine by the American Indians, who valued it highly. They boiled the root in water and drank the resulting beverage for a variety of conditions ranging from rheumatism, diseases of women, and debility to sore throat.[1] It was subsequently used, especially by eclectic physicians, for all these conditions but particularly for so-called uterine difficulties to stimulate the menstrual flow. Black cohosh was one of the principal ingredients in Lydia Pinkham's Vegetable Compound. Herbalists recommend it for all of the aforementioned ailments and also as an astringent, diuretic, alterative, antidiarrheal, cough suppressant, diaphoretic, and other uses.[2]

Scientific studies designed to identify specific physiological activities in the drug have not been numerous, and most have been carried out abroad. The long-suspected estrogenic effects, based on its use to stimulate menstruation, could not be verified in comprehensive experiments in mice reported in 1960.[3] Subsequent experiments have shown that a methanol extract of the herb contains substances that bind to estrogen receptors of rat uteri; the extract also causes a selective reduction in luteinizing hormone level in ovariectomized rats.[4] These results are interpreted to mean that black cohosh possesses some degree of estrogenic activity.

A steroidal triterpene derivative called actein, was found to lower blood pressure in rabbits and cats but not in dogs.[5,6] It produced no hypotensive effects in either normal or hypertensive human beings, although some peripheral vasodilation was observed.

Modern experience with black cohosh extracts dates to the mid-1950s. In Germany, gynecologists concerned with finding an alternative to hormone-replacement therapy, which by that time was showing unwanted side effects in a large number of patients, reported successful clinical experience in the treatment of menopausal symptoms with a black cohosh extract. By 1962, at least fourteen clinical reports, although not controlled clinical trials in the modern sense, involving over 1,500 patients were published in German. Practitioners reported efficacy in premenopausal and menopausal symptoms including reduction in hot flashes and improvement of "depressive moods."[7]

Since the 1980s, five clinical studies (although none with a double-blind design) have compared a black cohosh extract with placebo and/or estrogen replacement in the treatment of menopausal symptoms.[8] An open, multicenter study with data on 629 patients reported favorable results (in 80 percent of patients) after six to eight weeks of treatment. Improvements included relief of neurovegetative complaints such as hot flashes, sweating, headache, vertigo, palpitation, and tinnitus. Side effects (unspecified) were reported in 7 percent of patients, but they did not result in discontinuing therapy.[9]

A 1991 study confirmed an LH secretion inhibitory effect in both ovariectomized rats and in 110 menopausal women, demonstrating that the extract selectively suppresses luteinizing hormone secretion in menopausal women.[10]

A recent Japanese study reported positive effects of two Asian species, *C. heracleifolia* and *C. foetida,* on serum calcium and phosphate levels plus bone mineral density in rats. They concluded that ". . . Cimicifugae rhizome has potential in the treatment of osteoporosis, particulary in menopausal women."[11]

Black cohosh is prescribed in Europe for various conditions, including symptoms associated with premenstrual syndrome (PMS), dysmenorrhea, and menopause. Reported activities include an estrogen-like action, binding to estrogen receptors, and suppression of luteinizing hormone. Occasional stomach pain or intestinal discomfort has been reported.[12] Studies on mutagenicity, teratogenicity, and carcinogenicity have proven negative, and a six-month study on chronic toxicity in rats at about ninety times the human intake failed to prove deleterious.[13] Further studies on this useful herb are warranted.

REFERENCES

1. J. U. Lloyd. *Origin and History of all the Pharmacopeial Vegetable Drugs, Chemicals, and Preparations,* Volume 1. The Caxton Press, Cincinnati, Ohio, 1921, pp. 54-62.

2. R. C. Wren and R. W. Wren. *Potter's New Cyclopaedia of Botanical Drugs and Preparations,* New Edition. Health Science Press, Hengiscote, England, 1975, p. 89.

3. M. Siess and G. Seybold. *Arzneimittel-Forschung* 10: 514-520, 1960.

4. *Lawrence Review of Natural Products*, April, 1986.

5. E. Genazzani and L. Sorrentino. *Nature* 194: 544-545, 1962.

6. S. Corsano, G. Piancatelli, and L. Panizzi. *Gazzeta Chimica Italiana* 99: 915-932, 1969.

7. S. Foster. *Black Cohosh*-Cimificuga racemosa, *Botanical Series, No. 314.* American Botanical Council, Austin, Texas, 1998, 8 pp.

8. V. Schulz, R. Hänsel, and V. E. Tyler. *Rational Phytotherapy: A Physicians' Guide to Herbal Medicine,* Third Edition. Springer, Berlin, 1998, pp. 243-244.

9. H. Stolze. *Gyne* 3: 14-16, 1982.

10. E. M. Düker, L. Kopanski, H. Jarry, and W. Wuttke. *Planta Medica* 57: 420-424, 1991.

11. J. X. Li, S. Kadota, H. Y. Li, T. Miyahara, Y. W. Wu, H. Seto, and T. Namba. *Phytomedicine* 3(4): 379-385, 1996/1997.

12. *Bundesanzeiger.* January 5, 1989.

13. N. Beuscher. *Zeitschrift für Phytotherapi* 16: 301-310, 1995.

Blue Cohosh

One of the oldest indigenous American plant drugs is blue cohosh, otherwise known as papoose root or squaw root. It consists of the underground parts (roots and rhizomes) of *Caulophyllum thalictroides* (L.) Michx., a perennial herb, purple when young, that has a smooth stem, one to three feet in height, terminated by a panicle of yellowish green flowers. The mature plant is a peculiar bluish green color and bears dark blue fruits—hence, the name, blue cohosh. It is a member of the family Berberidaceae.[1]

The genus *Caulophyllum* contains five species, two from eastern North America [*C. thalictroides* and *C. giganteum* (Farw.) Loconte & Blackwell] and three from northeast Asia.[2] The rhizome of Asian species, *C. robustum* Maxim., has been used as a folk medicine to treat menstrual disorders.[3]

Blue cohosh was introduced into medicine in 1813 by Peter Smith, an "Indian herb doctor." It was said to be employed by the Indians for rheumatism, dropsy, colic, sore throat, cramp, hiccough, epilepsy, hysterics, inflammation of the uterus, etc. Subsequently, it gained a reputation as an antispasmodic, emmenagogue (menstrual flow stimulant), and parturifacient (inducer of labor), as well as a diuretic, diaphoretic, and expectorant. Modern herbals still recommend it for various female conditions, especially as a uterine stimulant, inducer of menstruation, and antispasmodic.[4]

The plant contains a number of alkaloids and glycosides, of which the alkaloid methylcytisine and the glycoside caulosaponin seem to contribute most of the physiological activity. Animal experiments have shown that the actions of methylcytisine resemble those of nicotine.[5] The compound elevates blood pressure and stimulates both respiration and intestinal motility. It is only about $\frac{1}{40}$ as toxic as nicotine. Blue cohosh's oxytocic (hastening childbirth) effects are apparently produced by the glycoside caulosaponin, a derivative of the triterpenoid saponin hederagenin.[6,7] Caulosaponin constricts the coronary blood vessels, thus exerting a toxic effect on cardiac muscle, and causes intestinal spasms in small animals.

In view of the presence of such relatively potent principles in the drug, blue cohosh cannot be dismissed as either inactive or harmless. The case for, or against, using it as self-medication, particularly to stimulate uterine contractions or to induce menstruation, probably rests on the advisability of using any self-selected drug for such purposes. The safety of such treatment is by no means certain. Kingsbury points out that the toxicity of English ivy is probably the result of the presence of a saponin glycoside derived, similar to caulosaponin, from hederagenin.[8] Discretion dictates that blue cohosh not be used for medical self-treatment.

REFERENCES

1. H. W. Felter and J. U. Lloyd. *King's American Dispensory,* Eighteenth Edition, Volume 1. The Ohio Valley Co., Cincinnati, Ohio, 1898, pp. 468-472.
2. H. Loconte and W. H. Blackwell. *Rhodora* 87: 463-469, 1985.
3. S. Foster. *East West Botanicals: Comparisons of Medicinal Plants Disjunct Between Eastern Asia and Eastern North America.* Ozark Beneficial Plant Project, Brixey, Missouri, 1986, pp. 12-13.
4. M. Castleman. *The Healing Herbs.* Rodale Press, Emmaus, Pennsylvania, 1991, pp. 82-84.
5. C. C. Scott and K. K. Chen. *Journal of Pharmacology and Experimental Therapeutics* 79: 334-339, 1943.
6. H. C. Ferguson and L. D. Edwards. *Journal of the American Pharmaceutical Association* (Scientific Edition) 43: 16-21, 1954.
7. J. McShefferty and J. B. Stenlake. *Journal of the Chemical Society* 2314-2316, 1956.
8. J. M. Kingsbury. *Poisonous Plants of the United States and Canada.* Prentice-Hall, Englewood Cliffs, New Jersey, 1964, pp. 371-372.

Boneset

Names of plants often reveal much information about them. They can also be misleading. There is little difficulty with the scientific name of boneset, *Eupatorium perfoliatum* L. The genus name of this member of the daisy family (Asteraceae) derives from Mithridates Eupator, ancient king of Pontus, who first used a closely related plant for medicinal purposes. The species designation, *perfoliatum,* refers to the manner in which the erect hairy stem of the hardy perennial herb, which attains a height of about five feet and is crowned with heads of white tubular florets, appears to perforate the center of the pairs of oppositely joined leaves. Boneset, the common name, is more likely to lead one astray since the plant was classically employed in the treatment of fevers, not to mend broken bones. However, when it is recognized that the old name for dengue was breakbone fever, the derivation becomes clear.[1]

American Indians introduced the use of boneset leaves and flowering tops to the early settlers for the treatment of colds, catarrh, influenza, rheumatism, and all kinds of fevers, including breakbone (dengue), intermittent (malaria), and lake (typhoid). To break up colds and flu, the drug is taken in the form of a hot tea to induce sweating and relieve the associated aches and pains. For loss of appetite, indigestion, and as a general bitter tonic, cold boneset infusion is recommended thirty minutes before meals.[2] In either case, the remedy is a bitter, astringent one with a nauseous taste. The hot version is much more likely to cause vomiting than the cold.[3]

Chemical studies have identified some of the constituents of boneset, which include various flavonoid pigments, sterols, and triterpenes.[4,5,6] Compounds with pronounced therapeutic virtues are generally absent. However, it has been reported that xyloglucurans from the polysaccharide fractions of aqueous extracts of boneset increased phagocytosis by a factor of 1 to 2.5 in the carbon clearance and granulocyte test, suggesting immunostimulating activity.[7] Eclectic physicians reported using boneset as an effective preventative and treatment for the "Spanish influenza" epidemic of 1918, as well as flu epidemics of the nineteenth century.[8]

The plant held official drug status in the United States from 1820 to 1950, even though it was rarely prescribed by physicians, at least during the latter part of that period.[9] Nevertheless, there is presently a revival of interest in the use of boneset among adherents to herbal medicine who employ it primarily to relieve fevers. Although safer and more effective treatments, such as common aspirin, certainly exist, it is comforting to know that the medical literature is essentially devoid of reports of adverse incidents attributed to boneset. Given the presence of potential immunostimulating polysaccharides, coupled with historical reports of efficacy in the prevention and treatment of influenza in the nineteenth and early twentieth centuries, this herb, relegated to historical obscurity, is deserving of closer scientific scrutiny.

REFERENCES

1. W. H. Hylton (Ed.). *The Rodale Herb Book.* Rodale Press Book Div., Emmaus, Pennsylvania, 1974, pp. 369-371.

2. E. Gibbons. *Stalking the Healthful Herbs,* Field Guide Edition. David McKay Co., New York, 1970, pp. 27-29.

3. L. Aikman. *Nature's Healing Arts: From Folk Medicine to Modern Drugs.* National Geographic Society, Washington, DC, 1977, pp. 40-42.

4. H. Wagner, M. A. Iyengar, L. Hörhammer, and W. Herz. *Phytochemistry* 11: 1504-1505, 1972.

5. W. Herz, S. Gibaja, S. V. Bhat, and A. Srinivasan. *Phytochemistry* 11: 2859-2863, 1972.

6. X. A. Domínguez, J. A. González Quintanilla, and M. Paulino Rojas. *Phytochemistry* 13: 673-674, 1974.

7. H. Wagner and A. Proksch. Immunostimulatory Drugs of Fungi and Higher Plants. In *Economic and Medicinal Plant Research,* Volume 1, H. Wagner, H. Hikino, and N. R. Farnsworth (Eds.). Academic Press, Orlando, Florida, 1985, pp. 134-135.

8. Lloyd Brothers, Pharmacists, Inc. *A Treatise on Eupatorium Perfoliatum.* Drug Treatise No. XXI. Lloyd Brothers Pharmacists, Inc., Cincinnati, Ohio, nd, 13 pp.

9. A. Osol and E. Farrar Jr. (Eds.). *The Dispensatory of the United States of America,* Twenty-Fourth Edition. J. B. Lippincott, Philadelphia, Pennsylvania, 1947, pp. 461-462.

Borage

Eating the leaves and flowers of borage, a hairy annual herb attaining a height of about two feet, has long been believed to confer both happiness and courage on the consumer. Indeed, Gerard tells us that such attributes of *Borago officinalis* L. (family Boraginaceae) were recorded as early as the first century A.D. when Pliny and Dioscorides observed that these plant parts added to wine made persons glad and merry.[1] In the second century A.D., Galen noted that the leaves possessed diuretic properties and also, when mixed with honey, were useful in treating throat irritations. These recommendations are still being repeated, nearly 2,000 years later, by modern herbalists.[2,3]

Fresh borage has a slightly salty taste, and its odor is reminiscent of cucumbers. It is sometimes eaten like spinach, either raw or cooked. Natural food enthusiast Euell Gibbons has provided recipes for borage syrup, borage candy, borage jam, and borage jelly.[4] Some people even drink borage tea.

But if you are expecting truly beneficial medicinal effects from this plant, try something else. Tests carried out in small animals found borage to be practically inert, except for a slight constipating effect attributable to its tannin content.[5] Tannin also accounts for the plant's astringent properties, whereas the presence of mucilage explains its mild expectorant action. The diuretic effects, if any, have been variously attributed to malic acid and to potassium nitrate.[6] Borage does contain appreciable concentrations of mineral salts. The seed oil contains about 20 percent of gamma-linolenic acid, a potentially valuable dietary supplement.[7]

As for the plant's ability to do away with sadness, dullness, and melancholy, we have to remember that the flowers and leaves were first soaked in wine which was then drunk to accomplish this purpose. Sufficient quantities of wine, without the borage, will achieve the same result—temporarily. And borage no more stimulates real courage than the potion in the square green bottle that the Wizard fed to the Cowardly Lion in *The Wizard of Oz.* But the belief has left us with an interesting old Latin verse:

Ego Borago gaudia semper ago.

Gerard renders this in the English of 1597 as:

I Borage bring alwaies courage.

In 1950, Hannig summarized our knowledge of the therapeutic utility of borage (in translation):

> In conclusion, the drug certainly will not enrich our materia medica. It may be numbered among the numerous tannin and mucilage containing drugs whose actions are nonspecific.[8]

Borage seed oil products are commonly sold in health food stores as an alternative source to evening primrose oil. The seeds contain up to 33 percent of a fixed oil comprised of 20 to 22 percent of gamma-linolenic acid. Little research supports its use. One study showed the oil may reduce reaction to stress in rats. A small study in ten patients found that the oil reduced stress by lowering systolic blood pressure and heart rate.[9]

In the mid-1980s, relatively low levels of a number of pyrrolizidine alkaloids (PAs), including lycopsamine, amabiline, and thesinine, were detected in various parts of the borage plant.[10] The unsaturated types (UPAs), such as amabaline, are notorious for their potential hepatotoxic and carcinogenic potential. Amabiline could not be detected in the seed oil down to 5 parts per million (ppm), but it has been noted that products containing levels as low as 1 ppm of such potent substances may prove harmful when taken internally for long periods of time.[11] In view of this, chronic consumption of either the borage plant or its seed oil should be carried out only under medical supervision, unless the products are certified free of UPAs.

REFERENCES

1. J. Gerard. *The Herball or Generall History of Plants.* John Norton, London, 1597, pp. 652-654.

2. R. C. Wren and R. W. Wren. *Potter's New Cyclopaedia of Botanical Drugs and Preparations,* New Edition. Health Science Press, Hengiscote, England, 1975, pp. 45-46.

3. W. H. Hylton (Ed.). *The Rodale Herb Book.* Rodale Press Book Div., Emmaus, Pennsylvania, 1975, pp. 371-375.

4. E. Gibbons. *Stalking the Healthful Herbs,* Field Guide Edition. David McKay Co., New York, 1970, pp. 53-57.

5. E. Hannig. *Die Pharmazie* 5: 35-40, 1950.

6. H. W. Felter and J. U. Lloyd. *King's American Dispensatory,* Eighteenth Edition, Volume 1. The Ohio Valley Co., Cincinnati, Ohio, 1898, p. 644.

7. D. V. C. Awang. *Canadian Pharmaceutical Journal* 123: 121-126, 1990.

8. E. Hannig, op. cit., pp. 35-40.

9. C. A. Newall, L. A. Anderson, and H. D. Phillipson. *Herbal Medicines: A Guide for Health-Care Professionals.* The Pharmaceutical Press, London, 1996, p. 49.

10. D. V. C. Awang, op. cit., pp. 121-126.

11. P. A. G. M. de Smet. *Canadian Pharmaceutical Journal* 124: 5, 1991.

Bran

Dietary fiber may be defined as all foods eaten by a monogastric (one stomach) animal that reach the large intestine essentially unchanged. This includes cellulose, the skeletal material of plant cell walls, and lignin, the rigid component of peach pits and nut shells, for instance. Together these materials are known as "crude fiber." Add to crude fiber such basically indigestible plant cell contents as gums, mucilages, pectic substances, hemicelluloses, and certain complex polysaccharides and the total is collectively referred to as "dietary fiber."[1]

A cheap and abundant source of dietary fiber is bran, the coarse outer coat or hull of the grain of wheat, *Triticum aestivum* L. (family Poaceae), separated from the meal or flour by sifting or bolting. Bran may also be obtained from other cereal grains, but then the source must be specified. Brans from oats, *Avena sativa* L., and from rice, *Oryza sativa* L., have received particular attention in recent years. The popularity of each has varied periodically, depending on the most recent findings of its utility as published in the medical press and disseminated to an anxious public by the media. It is too early to draw any final conclusions about the relative merits of the various products. Bran contains about 26.7 percent dietary fiber: by comparison, baked beans have 7.27 percent; boiled carrots, 3.7 percent, and whole peaches, 2.28 percent.[2] Lettuce, which many persons consider as real "roughage," contains only about 0.85 percent dietary fiber.[3]

Studies carried out in the early 1970s on rural black populations in South Africa showed that persons eating more than 50 grams of dietary fiber per day seemed to be relatively free from appendicitis and diseases of the colon (diverticulosis, polyps, or cancer) and to suffer less from ischemic heart disease (coronary artery disease), diabetes, and hiatus hernia than those persons eating more refined diets containing much smaller amounts of fiber. Interestingly, when the rural blacks moved to the cities and adopted Western habits, including diet, they began to suffer from the same Western diseases.

The beneficial effects of high dietary fiber foods such as bran on various gastrointestinal diseases, including cancer, are thought to come from (1) its ability to dilute, in the large amount of water held by the fiber, any carcinogens that might be present and (2) to decrease the contact time with potentially damaging substances, since the larger stool is expelled more rapidly. Passing stools more easily causes less straining and therefore reduces the frequency of hiatus hernia and hemorrhoids. Although the preventive effects of high-fiber diets on cardiovascular disease and diabetes are somewhat more speculative, these may result from the ability of specific polysaccharides to exert hypocholesteremic (lower cholesterol) and hypoglycemic (lower blood sugar) effects in humans.[4]

Another way of expressing these actions is to say that the fiber passes through the gut somewhat like a large wet sponge, absorbing and holding not only water and toxicants but such compounds as bile acids, which in turn might modify cholesterol metabolism. The sponge, due to its great bulk, also increases the size of the stool and decreases the emptying time of the colon.[5]

Bran is commercially available in a variety of forms, ranging from the crude material sold in bulk to compressed tablets. Probably the most popular sources in the United States are breakfast cereals; the whole bran types contain about 9 grams of dietary fiber per ounce (28.4 grams), which is approximately one-half cup. Of course, you should remember that bran is only one kind of dietary fiber and that many other varieties are present in such foods as legumes and fruits. Eating these is also desirable.

Although the advantages of a diet high in fiber are based on evidence from epidemiologic, experimental, and clinical studies,[6] it will be many years before all the hypotheses to account for these observations can be thoroughly tested. In the meantime, it is not surprising that opinions are wide-ranging on the amount of dietary fiber to be eaten daily by an adult. At present, the average American diet provides only 2 to 5 grams per day. Recommendations range from 10 grams[7] to as much as 60 grams.[8] An intermediate figure is probably more realistic, but this must be based largely on the individual's tolerance to increased bulk in the diet.

For best results, an increased intake in dietary fiber should be accompanied by a reduction in foods high in sugar, salt, and fat. Where possible, get the dietary fiber from whole-grain cereals and bread, potatoes, vegetables, and fruits. Supplement as necessary with bran. This may not result in "instant health," but there is a good possibility that it may reduce the frequency of certain chronic illnesses.

REFERENCES

1. A. I. Mendeloff. *New England Journal of Medicine* 297: 811-814, 1977.
2. V. E. Tyler, L. R. Brady, and J. E. Robbers. *Pharmacognosy,* Ninth Edition. Lea and Febiger, Philadelphia, Pennsylvania, 1988, p. 463.
3. R. G. Marks. *Drug Topics* 124(15): 20, 1980.
4. A. I. Mendeloff, op. cit., pp. 811-814.
5. P. Gunby. *Journal of the American Medical Association* 238: 1715-1716, 1977.
6. R. Adams and F. Murray. *Health Foods.* Larchmont Books, New York, 1975, pp. 33-39.
7. M. A. Krupp, M. J. Chatton, and L. M. Tierney Jr. *Current Medical Diagnosis and Treatment 1986.* Lange Medical Publications, Los Altos, California, 1986, pp. 810-811.
8. A. I. Mendeloff, op. cit., pp. 811-814.

Broom

The tops of *Cytisus scoparius* (L.) Link [*Sarothamnus scoparius* (L.) Koch.], a large, yellow-flowered shrub growing extensively along the Atlantic Coast and in the Pacific Northwest, are known variously as broom, broom tops, and Scotch broom. This member of the family Fabaceae was formerly used in medicine, primarily in the form of a fluid extract, for its diuretic, cathartic, and, in large doses, emetic properties. Broom contains up to about 1.5 percent of the alkaloid sparteine, once employed therapeutically to slow the pulse in cardiac disturbances; this use has now been discontinued in the United States. The plant also contains several other alkaloids and a number of simple amines that contribute to its physiological actions.[1]

In the 1970s, broom gained notoriety as a "legal" intoxicant or drug of abuse. Publications aimed at readers in the counterculture[2,3] recommended that the flowers be collected and aged about ten days in a sealed jar. The moldy, dried blossoms thus obtained are then pulverized, rolled in cigarette paper, and smoked like marijuana. One such cigarette is said to produce a feeling of relaxation and euphoria lasting about two hours. Smoking additional amounts, it is claimed, results in deeper relaxation and heightened color awareness, without any accompanying visual disturbances or hallucinations.

Although advocates maintain that such use of broom usually does not produce any undesirable effects except a mild headache, the practice has nothing to recommend it. Sparteine, the principal alkaloid in the plant, is contained in the various floral parts in concentrations ranging from 0.01 to 0.22 percent.[4] It is a volatile compound that is unquestionably present in the smoke inhaled. Sparteine has very pronounced physiological effects, slowing the heart (but not strengthening it) and stimulating uterine contractions. Because of the latter action, the alkaloid was formerly used to induce labor, an application found to be unsafe and now discontinued.

However, the greatest potential danger in smoking moldy broom flowers probably lies not in the flowers, but in the fungus infecting them. Studies have shown that almost all samples of illegally obtained marijuana tested were contaminated with pathogenic, inhalable *Aspergillus* species, which may sensitize the smoker.[5] Broom flowers purposely allowed to become moldy would almost certainly be similarly contaminated with such organisms and would thus present an appreciable risk to the user.

Broom does act as a diuretic as a result of having a flavone glycoside, scoparoside, present mostly in the flowers.[6] There are, however, safer and more effective drugs for this and all other actions attributed to the drug. Its employment as a mind-altering agent is potentially dangerous to the user in many ways and is definitely not recommended.

REFERENCES

1. A. Osol and G. E. Farrar Jr. (Eds.). *The Dispensatory of the United States of America,* Twenty-Fourth Edition. J. B. Lippincott, Philadelphia, Pennsylvania, 1947, p. 1579.

2. L. A. Young, L. G. Young, M. M. Klein, D. M. Klein, and D. Beyer. *Recreational Drugs.* Collier Books, New York, 1977, p. 42.

3. A Gottlieb. *Legal Highs*. 20th Century Alchemist, Manhattan Beach, California, 1973, pp. 6-7.

4. B. T. Cromwell. In *Modern Methods of Plant Analysis,* Volume 4, K. Paech and M. V. Tracey (Eds.). Springer-Verlag, Berlin, 1955, pp. 460-462.

5. S. L. Kagen, P. G. Sohnle, V. P. Kurup, and J. N. Fink. *New England Journal of Medicine* 304: 483-484, 1981.

6. E. Steinegger and R. Hänsel. *Lerbuch der Pharmakognosie*, Third Edition. Springer-Verlag, Berlin, 1972, pp. 291-292.

Buchu

Helmbold the Buchu King first came to my attention in a lecture I attended as an undergraduate student. My professor told his class how Henry T. Helmbold,[1] a patent medicine producer in New York City, introduced Helmbold's Compound Extract of Buchu in 1847 and advertised it widely as a cure for diabetes, gravel (kidney stones), inflammation of the kidneys, catarrh of the bladder, diseases arising from exposure or imprudence (venereal diseases), nervous diseases, prostration of the system, and the like . . . "from whatever cause originating and whether existing in either sex." As might be anticipated with a product such as that, he made a fortune and even aspired to the presidency of the United States!

Buchu consists of the dried leaves of three species of the genus *Barosma,* which are given common names based on the leaf shape: *B. betulina* (Thunb.) Bartl. & Wendl., commercially known as short buchu; *B. crenulata* (L.) Hook., called ovate buchu; *B. serratifolia* (Curt.) Willd., known as long buchu. All are obtained from low, white- or pink-flowered shrubs of the family Rutaceae, native to South Africa. Originally utilized as a drug by the Hottentots in that area, the leaves have been used as a household remedy for almost every known affliction. An alcoholic beverage, buchu brandy, is also widely distributed there.[2] The drug used to be official in *The National Formulary* and was rather widely employed as a urinary antiseptic and diuretic. Its use by physicians has been discontinued, but advocates of herbs continue to promote it for the same conditions that Helmbold recommended it more than 150 years ago.

Whatever therapeutic utility buchu may possess is due to its volatile oil, the principal active constituent of which is buchu camphor or diosphenol. Limonene, menthone, and pulegone are also major components. This accounts for the incorporation of buchu leaves in a large number of teas still sold in Europe for kidney and bladder conditions.[3] However, its diuretic and, especially, its antiseptic properties are relatively mild, and this must be kept in mind if one suffers from a condition that requires an especially effective drug. There is no reason, however, to question the safety of buchu, except in excessive amounts, due to possible toxicity of components in the essential oil.[4]

REFERENCES

1. H. W. Holcombe. *Patent Medicine Tax Stamps*. Quarterman Publications, Lawrence, Massachusetts, 1979, pp. 208-213.

2. H. S. Gentry. *Economic Botany* 15: 326-331, 1961.

3. E. Steinegger and R. Hänsel. *Lehrbuch der Pharmakognosie und Phytopharmazie,* Fourth Edition. Springer-Verlag, Berlin, 1988, pp. 695-696.

4. C. A. Newall, L. A. Anderson, and H. D. Phillipson. *Herbal Medicines: A Guide for Health-Care Professionals.* The Pharmaceutical Press, London, 1996, p. 51.

Burdock

Burdock consists of the dried, first-year root of great burdock, *Arctium lappa* L., or common burdock, *Arctium minus* (Hill) Bernh. (family Asteraceae). The former species is native to Europe but has been naturalized in the United States; the latter is the chief American source of the root. Both are large, coarse biennial herbs with hooked bracts or burrs that adhere to clothing or animal fur. *A. lappa* may attain a height of nine feet, but *A. minus* is limited to about five feet.

Recommended primarily as a blood purifier, burdock has also been used to treat various chronic skin conditions, including psoriasis and acne. It is also said to have diuretic and diaphoretic properties.[1] None of these purported effects has been verified by clinical trials, nor have chemical studies of the root revealed the presence of active principles that might account for any such effects. Burdock does contain large amounts of carbohydrate in the form of inulin, together with small amounts of volatile oil, fatty oil, sucrose, resin, tannin, and the like.[2] One interesting study did reveal the presence of some fourteen different polyacetylene compounds in the fresh root, two of which possessed bacteriostatic and fungistatic properties. However, only traces of them were found in the dried, commercial drug.[3] Mild antibiotic and cholagogic activity have been shown in root extracts. A study on the roots and leaves shows hypoglycemic activity and an increased carbohydrate tolerance in animal models. Antimutagenic activity has been demonstrated in animal models.[4]

Commercial samples of burdock are prone to adulteration with the root of belladonna, or deadly nightshade (*Atropa belladonna* L.). The roots of these plants closely resemble each other, and confusion often occurs when burdock is being harvested. Most burdock originates in Eastern Europe under circumstances in which strict quality control is lacking. At first, cases of atropine poisoning resulting from drinking burdock tea were attributed to the toxic effects of burdock itself.[5] Then, as additional cases were reported, both in the United States and in Europe, it became clear that the toxicity was the result of contamination with belladonna root.[6,7,8]

In view of this potential hazard, it seems reasonable to insist that marketers of burdock test their product and certify that it is free from belladonna (atropine) contamination prior to sale. Lacking the availability of a product of guaranteed quality, potential users with sufficient botanical knowledge will probably wish to collect the herb themselves.

However, despite its long use as a folkloric remedy, no solid evidence exists that burdock exhibits any useful therapeutic activity. The young leaves may be eaten as greens, and the fresh root may have some antimicrobial properties, but the commercial dried root is lacking in confirmed medicinal value.

REFERENCES

1. J. C. Torke. *Herbalist* 2(4): 124-125, 1977.

2. H. A. Hoppe. *Drogenkunde,* Eighth Edition, Volume 1. Walter de Gruyter, Berlin, 1975, pp. 102-104.

3. K. E. Schulte, G. Rücker, and R. Boehme. *Arzneimittel-Forschung* 17: 829-833, 1967.

4. F. Chandler and F. Osborne. *Canadian Pharmaceutical Journal* 130(5): 46-49, 1997.

5. P. D. Bryson, A. S. Watanabe, B. H. Rumack, and R. C. Murphy. *Journal of the American Medical Association* 239: 2157, 1978.

6. P. D. Bryson. *Journal of the American Medical Association* 240: 1586, 1978.

7. Anon. *Deutsche Apotheker Zeitung* 124: 390, 1984.

8. P. M. Rhoads, T. G. Tong, W. Banner Jr., and R. Anderson. *Journal of Toxicology. Clinical Toxicology* 22: 581-584, 1985.

Butcher's-Broom

Broom is one of those indefinite common names that tends to make the field of herbal nomenclature such a difficult one. The name was originally applied to several plants whose tough stems and rigid leaves made them useful for sweeping up debris. Used without a qualifying adjective, broom refers to the previously discussed *Cytisus scoparius* L., a common roadside plant in the Pacific Northwest, distinguishable by its showy yellow flowers. Spanish broom or gorse (*Spartium junceum* L.) is another yellow-flowered leguminous shrub that flourishes in parts of California. Although both plants have been used in folk medicine, neither is the so-called butcher's-broom that is, so to speak, "sweeping the country" at the present time.

Butcher's-broom, also known as box holly or knee holly, is a fairly common, short evergreen shrub (*Ruscus aculeatus* L.) of the family Liliaceae, native throughout the Mediterranean region from the Azores to Iran. It, too, has a long history of use in herbal medicine. As early as the first century, Dioscorides recommended butcher's-broom as a laxative and diuretic. The seventeenth-century apothecary-astrologer Nicholas Culpeper suggested that a decoction of the root be drunk and a poultice of the berries and leaves applied to facilitate the knitting of broken bones. However, the drug never became popular in either Europe or the United States and was seldom mentioned in standard references on drugs.[1]

Then, during the 1950s, French investigators showed that an alcoholic extract of butcher's-broom rhizomes (underground stems) produced vasoconstriction (blood vessel narrowing) in test animals.[2] Further studies identified the active principles as a mixture of steroidal saponins, the two main ones being identified as ruscogenin and neoruscogenin.[3,4] They apparently produce their vasoconstrictive effects by direct activation of α-adrenergic receptors.[5] Japanese researchers have isolated twelve steroidal saponins, including seven new ones, two of which have cytostatic activity on leukemia HL-60 cells.[6]

Limited clinical trials in humans have, in general, provided support for the effectiveness of the drug in venous disorders.[7,8,9,10] In addition to its vasoconstrictive effects, the extract was demonstrated to have anti-inflammatory properties.

Such studies convinced certain European drug manufacturers that butcher's-broom extract is superior to some of the conventional plant remedies, such as extracts of horse chestnut and witch hazel, which are marketed for their supposed beneficial effects on venous circulation. Consequently, they have made extracts of butcher's-broom commercially available in capsule form to treat circulatory problems of the legs and as an ointment or suppository to relieve the symptoms of hemorrhoids.[11]

Capsules containing 75 mg of butcher's-broom and 2 mg of rosemary oil are now being sold in the United States, mainly through health food stores. One such product is being advertised as "a proven European herbal formula—said to improve circulation in the legs," while another is being promoted with the claim that "millions of Europeans report it works wonders—particularly for women who often complain about a 'heavy feeling' in the legs." The ads also state that butcher's-broom is "rare" or "hard-to-find"—which is not true.[12]

Although there may be some basis for cautious optimism concerning butcher's-broom as a potentially useful drug, would-be consumers should recognize that manufacturers of butcher's-broom products have never presented proof of safety and efficacy to the Food and Drug Administration and that therapeutic claims for these products are therefore illegal. Moreover, self-diagnosis and self-treatment of circulatory disorders, or any other potentially serious health problem, are certainly inadvisable.

REFERENCES

1. I. Müller. *Deutsche Apotheke -Zeitung* 113: 1370-1375, 1973.

2. P. Caujolle, P. Mériel, and E. Stanislas. *Annales Pharmaceutiques Francaises* 11: 109-120, 1953.

3. P. H. List and L. Hörhammer (Eds.). *Hagers Handbuch der Pharmazeutischen Praxis*, Fourth Edition, Volume 6B. Springer-Verlag, Berlin, 1979, pp. 200-201.

4. C. Sannie and H. Lapin. *Bulletin de la Societe Chimique De France*: 1237-1241, 1957.

5. G. Rubanyi, G. Marcelon, and P. M. Vanhoutte. *General Pharmacology* 15: 431-434, 1984.

6. Y. Mimaki, M. Kuroda, A. Kameyama, A. Yokosuka, and Y. Sashida. *Chemical and Pharmaceutical Bulletin* (Tokyo) 46(2): 298-303, 1998.

7. N. Weindorf and U. Schultz-Ehrenburg. *Zeitschrift für Hautkrankheiten* 62: 28-30, 35-38, 1987.

8. R. Cappelli, M. Nicora, and T. DiPerri. *Drugs Under Experimental and Clinical Research* 14: 277-284, 1988.

9. D. Berg. *Fortschritte der Medizine* (Munchen) 110(3): 67-68, 71-72, 1992.

10. D. Berg. *Fortschritte der Medizine* (Munchen) 108(24): 473-476, 1990.

11. R. F. Weiss. *Lehrbuch der Phytotherapie,* Fifth Edition. Hippokrates Verglag, Stuttgart, Germany, 1982, pp. 142-143.

12. *Better Nutrition* 44(1): 17, 1984.

Caffeine-Containing Plants

Coffee, though a useful medicine, if drank constantly, will at length induce a decay of health, and hectic fever.

Jesse Torrey
The Moral Instructor
Pt. IV, Sect. II, Ch. 10

Six caffeine-containing plants are more widely used by humankind, primarily as beverages, than all the other herbal materials put together. These ubiquitous products include:[1]

Coffee—the dried ripe seed of *Coffea arabica* L. or other species of *Coffea* (family Rubiaceae), small evergreen trees or large shrubs cultivated extensively in tropical areas of the world. Deprived of most of the seed coat, the so-called bean is roasted until it becomes dark brown in color and develops a characteristic aroma. Coffee beans contain 1 to 2 percent caffeine.

Tea—The prepared leaves and leaf buds of *Camellia sinensis* (L.) O. Kuntze (family Theaceae), a large shrub with evergreen leaves native to eastern Asia and extensively cultivated there. Black tea is prepared by an initial slow drying of the fresh leaves that allows them to begin to ferment. For green tea, a less popular beverage in the United States, the leaves are quickly dried. Because of these different methods of preparation and the many different varieties of the cultivated plant, the average caffeine content of tea ranges widely, from about 1 to more than 4 percent.

Kola (Cola or Kolanuts)—the dried cotyledons (seed leaves) of *Cola nitida* (Vent.) Schott & Endl., or of other species of *Cola* (family Sterculiaceae). The plants are large trees native to West Africa, where much of our commercial supply still comes from, but are also cultivated in the West Indies and other tropical lands. Kola contains up to about 3 percent caffeine.

Cacao (Cocoa)—the roasted seeds of *Theobroma cacao* L. (family Sterculiaceae), a relatively tall tree native to Mexico but now widely cultivated in the tropics. With processing, cacao beans (also called cocoa beans) yield chocolate and all its related products such as breakfast cocoa and cacao (cocoa) butter. Cacao contains between 0.07 and 0.36 percent caffeine. IMPORTANT! Do not confuse cacao or cocoa, which are identical, with coca, the source of cocaine, or with coconut, the palm that yields copra and coconut oil. Cacao (cocoa), coca, and coconut are three entirely different plants, even if the old Marx Brothers' movie with its misspelled title *The Cocoanuts* did thoroughly confuse the issue.

Guarana—a dried paste made chiefly from the crushed seed of *Paullinia cupana* H.B.K. (family Sapindaceae), a climbing shrub native to Brazil and Uruguay. The natives of those countries prepare a hot beverage from guarana, and it is employed as an ingredient in carbonated beverages marketed by the Coca-Cola Company in Brazil and the Pepsi-Cola company in the United States. Guarana has a relatively high caffeine content, ranging from 2.5 to 5 percent and averaging about 3.5 percent.

Maté—the dried leaves of *Ilex paraguariensis* St. Hill (family Aquifoliaceae), a small tree or shrub growing wild in Paraguay, Brazil, and other South American countries. Containing up to 2 percent caffeine, maté, or Paraguay tea as it is frequently called, is widely used in South America in the preparation of a tealike beverage. Two related species of *Ilex*, *I. cassine* L. and *I. vomitoria* Ait., known as yaupon hollies among other names, native to the southeastern United States, were used as ceremonial beverages by Indian groups of the Southeast. An extract-like tea of the leaves that induced vomiting was drunk at councils and before going into battle. Ironically, these North

American caffeine-containing *Ilex* species have never been commercially developed. Perhaps if William Aiton (1731-1793) had chosen a specific epithet for yaupon other than "vomitoria," it may not have been relegated to historical obscurity.

All of these caffeine-containing plants (with the exception of *Ilex cassine* and *I. vomitoria*) are listed on the Food and Drug Administration's "Generally Recognized as Safe" (GRAS) list, indicating their acceptability as food additives.[2] Some of them, such as guarana, as well as the pure caffeine obtained from them, are sold in tablet form. Caffeine is also added to "cola" and "pepper"-type drinks, baked goods, frozen dairy desserts, gelatin puddings and fillings, soft candy, and the like. Several companies produce caffeine-containing tablets or capsules in forms resembling those used for the stimulant prescription pharmaceuticals commonly known as "uppers."[3] Caffeine is also utilized in so-called designer foods; one of them, WOW, is said to contain "enough caffeine to resurrect King Tut."[4]

The important information for the potential consumer of a caffeine-containing plant material is not how much active constituent is present in the crude drug but how much is in the consumed product. Here are approximate amounts of caffeine in some commonly used foods and beverages:[5,6,7]

Cup (6 oz) of boiled coffee	100 mg
Cup (6 oz) of instant coffee	65 mg
Cup (6 oz) of tea	10-50 mg
Bottle or can (12 oz) of cola beverage	50 mg
Cup (6 oz) of breakfast cocoa	13 mg
Bar (1 oz) of milk chocolate	6 mg
Tablet of caffeine (proprietary product)	100-200 mg
Tablet (800 mg) of guarana	30 mg
Cup (6 oz) of maté	25-50 mg

Note that the method of preparation is more important than the caffeine content of the original plant material in determining the concentration in the final product. Tea, which may contain 4 percent caffeine in its leaves, seldom has more than 50 mg per cup because it is prepared as an infusion (boiling water is added to the leaves and allowed to cool). Coffee, on the other hand, contains only 1 to 2 percent caffeine but is usually prepared as a decoction, that is, placed in cold water which is then slowly heated to boiling. Thus, a cup of coffee contains about twice as much caffeine (100 mg) as the same quantity of tea. Of course, the exact amounts vary appreciably because of the different quantities of plant material used by different preparers.

Caffeine is the principal physiologically active constituent in all of these plants, but to alert those concerned about their health, many contain other constituents that may be harmful for long-term usage. For example, there is evidence indicating that the condensed catechin tannin of tea is linked to high rates of esophageal cancer in areas where tea is consumed in large quantities.[8] Incidentally, this effect apparently may be overcome by consuming the tea along with milk, thereby binding the tannin and preventing its deleterious effect.

Caffeine is an effective stimulant of the central nervous system but, especially in large amounts, produces many undesirable side effects—from nervousness and insomnia to rapid and irregular heartbeats, elevated blood sugar and cholesterol levels, excess stomach acid, and heartburn. It is definitely a teratogen (produces deformed fetuses) in rats; the FDA has advised practitioners to counsel patients who are or may become pregnant to avoid or limit consumption of foods and drugs containing caffeine.[9]

This is sound advice, for based on our incomplete knowledge of side reactions to caffeine, prudent use seems desirable for all consumers, male and female. The problem is particularly significant in children because the effects of caffeine are related to body weight of the consumer. Nonpregnant adults should probably limit their consumption of caffeine to no more than 250 mg per day.[10] Pregnant women and children should be even more conservative. And moderation in the use of caffeine-containing products should be the watchword for all.

REFERENCES

1. V. E. Tyler, L. R. Brady, and J. E. Robbers. *Pharmacognosy,* Ninth Edition. Lea and Febiger, Philadelphia, Pennsylvania, 1988, pp. 244-248.

2. *Herbs on the GRAS List.* Compiled by Herb Research Foundation from Code of Federal Regulations, Title 21, Parts 172, 182, 184, and 186, July 16, 1984.

3. *High Times* 188: 9, 11, 65, 1991.

4. D. Kushner. *High Times* 187: 25, 1991.
5. A. Salsedo. *Whole Foods* 2(12): 20-31, 1979.
6. J. Eddington (Ed.). *Update: Cancer Research & Training Newsletter* (Purdue University Cancer Center) 5(8): 2, 1980.
7. *FDA Drug Bulletin* 10(3): 19-20, 1980.
8. J. F. Morton. *Science* 204: 909, 1979.
9. *FDA Drug Bulletin* 10(3): 19-20, 1980.
10. J. Eddington (Ed.), op. cit., p. 2.

Calamus

Known since biblical times, the aromatic rhizome (underground stem) of *Acorus calamus* L. is commonly referred to as calamus or sweet flag. It has been taken over the centuries as a remedy for various sorts of digestive upsets and colic, especially in children. The plant is a perennial herb of the family Araceae, commonly found in moist habitats such as the banks of ponds or streams and in swamps throughout North America, Europe, and Asia. In appearance, the leaves resembles the iris. Long placed in the Araceae (arum family), some plant scientists argue for its placement in a separate family designated the Acoraceae (acorus family).[1]

Modern writers on herbs recommend an infusion of the rhizome for fevers and dyspepsia; chewing the rhizome to ease digestion and to clear the voice; and using the powdered material as a substitute for various spices in cooking.[2] Some persons greatly enjoy its flavor. As Brer Rabbit put it, "I done got so now dat I can't eat no chicken 'ceppin she's seasoned up wid calamus root." Calamus was once used in this country as a flavoring agent in a variety of commercial products, ranging from tooth powders and tonics to beer and bitters. Candied sweet flag root was available into the 1960s, manufactured at the Sabbathday Lake, Maine, Shaker Community. Various species or varieties of this semiaquatic plant group from temperate regions of the Americas and Eurasia have a rich ethnobotanical history, encompassing numerous uses and cultures throughout their range. General ethnobotanical attributes include use as a stomachic, diaphoretic, diuretic, emmenagogue, appetite stimulant, and antispasmodic, among dozens of additional uses.[3]

The volatile oil responsible for the drug's characteristic odor and taste occurs in amounts ranging from 1.5 to more than 3.5 percent. Unfortunately, feeding studies conducted over thirty years ago established that β-asarone (*cis*-isoasarone), a major constituent in certain calamus oils, produced malignant tumors in the duodenal region of rats. Since then, use of calamus as a food or food additive has been banned, at least in the United States.[4]

Subsequent investigations have now shown that there are actually four different drug types of calamus, each originating from a different variety of *Acorus calamus* growing in different geographical areas of the world. Drug type I is found in North America, and its oil is isoasarone free. Drug type II is produced in western Europe from plants originating in eastern Europe; its volatile oil usually contains less than 10 percent isoasarone. Drug types III and IV are varieties whose volatile oils may contain as much as 96 percent *cis*-isoasarone.[5]

Botanical material from North America (drug type I) formerly designated *A. calamus* or *A. calamus* var. *americanus* (Raf.) H.D. Wulff is now recognized as a distinct species, *A. americanus* (Raf.) Raf., which is diploid. The typical European material is a sterile triploid.

Pharmacological tests have now shown that the isoasarone-free oil of drug type I (*A. americanus*) has an even more effective spasmolytic (antispasmodic) activity than the isoasarone-rich oil of drug type IV or the isoasarone-poor oil of drug type II.[6] Such results suggest that North American (type I) calamus is an effective herbal remedy for dyspepsia and similar conditions for which its antispasmodic effect may produce some relief. The identity of the constituents in the volatile oil that are responsible for this effect remain to be established. Although the absolute safety of *A. americanus* (type I calamus) has yet to be proven by extensive clinical tests, it is at least free of the carcinogenic isoasarone that renders the other drug types unsuitable for medicinal use. Certainly, more research is necessary.

REFERENCES

1. M. H. Grayum. *Taxon* 36: 723, 1987.
2. W. H. Hylton (Ed.). *The Rodale Herb Book*. Rodale Press Book Div. Emmaus, Pennsylvania, 1976, pp. 380-381.
3. T. J. Motley. *Economic Botany* 48(4): 397-412, 1994.
4. J. M. Taylor, W. I. Jones, E. C. Hagan, M. A. Gross, D. A. Davis, and E. L. Cook. *Toxicology and Applied Pharmacology* 10: 405, 1967.
5. K. Keller and E. Stahl. *Deutsche Apotheker Zeitung* 122: 2463-2466, 1982.
6. K. Keller, K. P. Odenthal, and E. Leng-Peschlow. *Planta Medica* 1: 6-9, 1985.

Calendula

The ligulate florets, commonly (but erroneously) referred to as flower petals, of *Calendula officinalis* L. have been used in medicine since the very earliest times. The plant, a member of the family Asteraceae, is a common cultivated ornamental, also referred to as pot marigold or garden marigold. It should not be confused with members of the genus *Tagetes,* also widely grown as garden ornamentals and commonly called marigolds. During its long history, calendula has been administered internally for a variety of ailments, including spasms, fevers, suppressed menstruation, and cancer. Its chief use, however, was as a local application to help heal and prevent infection of lacerated wounds.[1] Modern herbalists recommend it in the form of a tincture, infusion, or ointment to heal a variety of skin conditions ranging from chapped hands to open wounds.[2]

A large number of chemical studies of calendula flowers have been carried out, especially in Europe, without revealing any principles that are unique, or even outstanding, in their physiological properties. A volatile oil, bitter principles, carotenoids, mucilage, resin, polysaccharides, plant acids, various alcohols, saponins and other glycosides, and sterols are all present. Many of the individual constituents in these general groups have been identified.[3] The active principles responsible for calendula's purported anti-inflammatory and wound-healing properties when applied locally to the skin or mucous membranes remained unknown until the mid-1980s.[4] Although some anti-inflammatory activity has been attributed to both polysaccharides and saponins, they are not present in lipophilic extracts of the drug. Recently, the faradiol monoester of a purified triterpenoid fraction of the flowers (in a CO_2 extract) has been shown to play an important role in anti-inflammatory activity.[5] A recent study on a freeze-dried extract of the flowers showed that water-soluble flavonoids (or other components) increased the rate of neovascularization and induced deposition of hyaluronan, a major component of extracellular matrix, associated with the formation, alignment, and migration of newly formed capillaries. This research begins to provide some scientific evidence to confirm the drug's long-standing traditional use in wound healing.[6]

The carotenoid pigments possess some utility as coloring agents in cosmetics, and the volatile oil is a useful ingredient in perfume, but none of the other known components has medicinal properties that are superior to other available remedies. Calendula is apparently nontoxic and, in an ointment, will be colored sufficiently to delineate a wound or other skin condition where applied. In this regard, it is probably as useful as Mercurochrome.

REFERENCES

1. H. W. Felter and J. U. Lloyd. *King's American Dispensatory*, Eighteenth Edition. Volume 1. The Ohio Valley Co., Cincinnati, Ohio, 1898, pp. 401-403.

2. D. D. Buchman. *Herbalist* 2(3): 80-84, 1977.

3. P. H. List and L. Hörhammer (Eds.). *Hagers Handbuch der Pharmazeutischen Praxis,* Fourth Edition. Volume 3. Springer-Verlag, Berlin, 1972, pp. 603-608.

4. M. Wichtl (Ed.). *Teedrogen.* Wissenschaftliche Verlagsgesellschaft mbH, Stuttgart, Germany, 1984, pp. 274-276.

5. R. Della Loggia, A. Tuburo, S. Sosa, H. Becker, St. Sasar, and O. Isaac. *Planta Medica* 60: 516-520, 1994.

6. K. F. M. Patrick, S. Kumar, P.A.D. Edwardson, and J. J. Hutchinson. *Phytomedicine* 3(1): 11-18, 1996.

Canaigre

Canaigre, the root of *Rumex hymenosepalus* Torr., was marketed in the late 1970s under such coined names of modern vintage as wild red American ginseng and wild red desert ginseng. The plant, a member of the family Polygonaceae, native to the deserts of the southwestern United States and Mexico, actually bears no relationship either botanically or in its active principles to ginseng, which belongs to an entirely different family, the Araliaceae.

Promotional literature on canaigre indicates that the drug was recommended in old herbals for a large number of maladies, ranging from lack of vitality to leprosy.[1] Unfortunately, the authors of such statements somehow fail to include references, and an inspection of herbal literature does not substantiate this claim. For example, J. M. Nickell's comprehensive listing of some 2,500 botanical remedies in 1911 omits the plant entirely.[2] *King's American Dispensatory* (1900), a comprehensive eclectic compendium devoting 2,172 closely written pages to plant remedies, dedicates eight lines of fine print to canaigre, stating that because of its high tannin content, it was used for tanning and dyeing by the Indians.[3] Not a single word therein notes any medicinal use. Voelcker, however, mentions in 1876 that the natives of Mexico used the root as an astringent.[4] In the Southwest, it was known as the "Indian tan plant." In 1896, the state of Texas published an agricultural bulletin promoting its potential as a new crop for the tanning industry.[5]

It is obvious that the attempt to promote canaigre as a kind of American ginseng is a recent-day deceptive practice, probably due to the high prices now commanded by ginseng. Canaigre does not contain any of the active panaxoside-like saponin glycosides responsible for ginseng's physiological activities. It does contain 18 to 25 percent or more of tannin and smaller amounts of anthraquinones, as well as other constituents such as starch and resin.[6]

Recognizing this attempt to substitute a relatively common, essentially worthless plant for a more valuable commodity, the Herb

Trade Association adopted a policy statement in 1979 "that any herb products consisting in whole or part of *Rumex hymenosepalus* should not be labeled as containing 'ginseng.'"[7] Nevertheless, canaigre's synonym, red American ginseng, still exists in the herbal literature of the 1970s and 1980s and may cause the unsophisticated reader to confuse it with authentic red ginseng or even American ginseng. They are by no means similar. In the herbal annals of the latter twentieth century, the canigre story will survive as a classic case of adulterated foolery.

Canaigre may be a useful material for tanning leather and dyeing wool, but it has no place in therapeutics. Indeed, because of its high tannin content, the root may have considerable carcinogenic potential. Rational people will avoid using it or any capsules or extracts prepared from it.

You can't make a silk purse out of a sow's ear.

Jonathan Swift
Polite Conversation

REFERENCES

1. D. L. Poulson. *Herbalist* 3(1): 16-17, 1978.
2. J. M. Nickell. *J. M. Nickell's Botanical Ready Reference.* Murray and Nickell Mfg. Co., Chicago, Illinois, 1911.
3. H. W. Felter and J. U. Lloyd. *King's American Dispensatory,* Eighteenth Edition, Volume 2. The Ohio Valley Company, Cincinnati, Ohio, 1900, p. 1685.
4. Herb Trade Association. Policy Statement #1—Canaigre, May, 1979.
5. R. F. G. Voelcker. *American Journal of Pharmacy* 48: 49-51, 1876.
6. C. Wehmer. *Die Pflanzenstoffe,* Second Edition, Volume 1. Verlag von Gustav Fischer, Jena, Germany, 1929, p. 275.
7. Herb Trade Association. Policy Statement #1—Canaigre, May, 1979.

Capsicum

Capsicum, cayenne pepper or chili pepper, is also referred to merely as red pepper. It consists of the dried ripe fruit of *Capsicum frutescens* L., *Capsicum annuum* L., or a large number of hybrids or varieties of these species of the family Solanaceae. For centuries, these plants have been highly valued as spices, and the extensive cultivation carried out over that period of time has resulted in peppers widely differing from one another in size, shape, and pungency. The labeling of commercial samples is really meaningful only if the variety is specifically designated.

Applied externally, capsicum is a rubefacient, that is, an agent that reddens the skin, thereby producing a counterirritant effect. Internally, it is valued as a stomachic, carminative, and gastrointestinal stimulant.[2] Numerous popular herbals tout its use as a circulatory stimulant, which can be traced back to its introduction to popular medicine by Samuel Thomson (1769-1843), who promoted it as his "Number 2" remedy (second to *Lobelia inflata*) in his "Thomsonian System," which advocated "cleansing" the patient with steaming, sweating, vomiting, and stimulants.[3]

All of these activities attributed to capsicum depend upon the presence in capsicum of a compound known as capsaicin, which, together with two closely related principles, is responsible for the pungency of the fruit. In addition, capsicum serves as a relatively good source of vitamin C.

In addition to its proven effectiveness in various gastrointestinal conditions, some claims have been made regarding the utility of capsicum in reducing blood cholesterol levels and decreasing the tendency of the blood to clot.[4,5] At present, supporting evidence for such claims is insubstantial. However, the local application of capsaicin ointment to treat chronic pain due to herpes zoster (shingles), trigeminal neuralgia, or surgical trauma (stump pain following amputation) is now well verified. Such application has even given some preliminary indications of effectiveness in the treatment of cluster headache.

Capsaicin functions in such cases by causing the depletion of substance P, the compound mediating the transmission of pain impulses from the peripheral nerves to the spinal cord. Thus, even though the condition causing the pain is still present, no perception of it reaches the brain. The initial depletion of substance P requires about three days, so even regular application of the ointment does not bring immediate relief.[6]

As an external counterirritant, capsicum is quite effective. Its action is fairly long lasting, and it does not blister the skin.[7] However, care must be taken not to get it into the eye, or extreme discomfort may result. Capsicum-containing mixtures intended to be placed in the stockings to keep the feet warm on cold days are currently marketed. Although they may be modestly effective, the temporary warmth they provide is offset by the risk that when used by children, some of the pepper may accidentally get into their eyes, resulting in considerable pain.

Persons who pick or otherwise handle quantities of hot peppers recognize that the pungent principle capsaicin is essentially insoluble in cold water and only slightly soluble in hot water. Traces remaining on the hands may be transferred inadvertently to sensitive mucous membranes even several hours after contact. The capsaicin may be removed from the affected part of the anatomy by bathing in vinegar, but, of course, this should not be applied in or around the eye.[8]

Internally, there are great individual variations in the sensitivity of persons to capsicum. The quantity that would prove to be a useful stomachic or digestive aid in one person might be very irritating, even upsetting, to another. Caution in the use of this irritating product is certainly advisable.

REFERENCES

1. V. E. Tyler, L. R. Brady, and J. E. Robbers. *Pharmacognosy*, Ninth Edition. Lea and Febiger, Philadelphia, Pennsylvania, 1988, pp. 148-150.

2. R. A. Locock. *Canadian Pharmaceutical Journal* 118: 516-519, 1985.

3. S. Thomson: *New Guide to Health; or, Botanical Family Physician.* J. Q. Adams, Boston, Massachusetts, 1835.

4. M. Castleman. *Medical Self-Care* 52: 68-69, 1989.
5. S. Visudhiphan, S. Poolsuppasit, O. Piboonnukarintr, and S. Tumliang. *American Journal of Clinical Nutrition* 35: 1452-1458, 1982.
6. *Lawrence Review of Natural Products,* November, 1989.
7. T. Sollmann. *A Manual of Pharmacology,* Seventh Edition. W. B. Saunders Company, Philadelphia, Pennsylvania, 1948, p. 137.
8. T. P. Vogl. *New England Journal of Medicine* 306: 178, 1982.

Catnip

Cats are said to be very fond of it.

It seems strange that the relationship between cats and catnip should have been regarded with such skepticism by an authority on drugs of such stature as *The United States Dispensatory,* Second Edition, published in 1834. Not until the seventeenth edition, published in 1896, was the sentence changed to: "Cats are very fond of it . . . ." Apparently, after more than sixty years, one of the authors of this respected reference finally observed the not uncommon interaction between a cat and the plant whose various names, catnip, catnep, and catmint, all suggest fairly close affinity.

The dried leaves and flowering tops of the common, wayside, perennial plant *Nepeta cataria* L. (family Lamiaceae) contain a volatile oil that is indeed extremely attractive to cats, causing them to cavort playfully while attempting to saturate their entire bodies with the plant's distinctive aroma.[1] Catnip was once rather widely used in human medicine, primarily as a carminative or digestive aid and as a tonic. The hot tea taken at bedtime has also been recommended as a sleep aid. But Gibbons gives his opinion, and it is possibly factual, that most catnip tea enthusiasts drink it simply because they like the taste of this pleasant beverage.[2]

In more recent times, catnip has been promoted by certain members of the counterculture as a psychedelic drug, said to produce a sense of well-being or euphoria when smoked like tobacco or marijuana. Unfortunately, the physicians who first described this use in an article in *The Journal of the American Medical Association* confused the identity of the drug, erroneously labeling a drawing of a marijuana plant as catnip and vice versa.[3] Some 1,612 letters pointing out this error were received by the editor, and one writer commented, "Perhaps one reason for his patients getting high on catnip is their lack of botanical knowledge."[4]

It is an unfortunate fact that once an erroneous statement has appeared in print, it is almost impossible to eradicate it. Catnip now is listed in practically all books devoted to drugs of abuse as a mild intoxicant. One of the more cautious of these begins its discussion of the plant with the statement, "Does it or doesn't it?"[5] The book then goes on to indicate that a debate has raged for years among potheads as to whether one can actually get "high" by using catnip. Any drug whose mind-altering effects are as questionable as this one is scarcely worth considering for that purpose.

Recently, catnip was reported to alter human consciousness in the form of an adverse reaction. It comes from an emergency room report, in which a nineteen-month-old male child was brought to a hospital because his mother thought he "looked drugged." A day before the child was brought to the emergency room, he had been eating raisins soaked in a commercial catnip tea, which had sat in a refrigerator for three weeks. The child was found to be chewing on the bag of catnip tea, which was taken away from him. After several hours passed, the child developed a stomachache and became irritable. The next morning the mother took him to the hospital. Various tests failed to show any medical problem that could account for the child's lethargy and apparent hypnotic condition. After two and a half days in the hospital, the child had a bowel movement, passing the raisins and catnip tea leaves. Soon after, he returned to a normal state of consciousness and was sent home from the hospital. This may be the first documented case of catnip toxicity. The next obvious question might be "Was it really *Nepeta cataria* in the tea bag labeled as catnip?"[6]

The plant does contain a volatile oil, the odor of which causes a series of characteristic stimulatory responses in *cats*. The active principle responsible for these effects is a compound designated *cis-trans*-nepetalactone; it is a major component, constituting 70 to 99 percent of the volatile oil. Interestingly, it produces its effects in cats only when it is smelled, not when it is administered orally.[7] Although there is now no scientific evidence to support the sedative activity of catnip tea, many people do drink a cup of it at bedtime in the belief that it will ensure a good night's sleep. Besides, it is relatively inexpensive, it tastes good, and no harmful effects from using moderate amounts of freshly prepared tea have been reported. What more can one ask of a beverage?

REFERENCES

1. A. O. Tucker and S. S. Tucker. *Economic Botany* 42: 214-231, 1988.

2. E. Gibbons. *Stalking the Healthful Herbs,* Field Guide Edition. David McKay Co., New York, 1970, p. 84.

3. B. Jackson and A. Reed. *Journal of the American Medical Association* 207: 1349-1350, 1969.

4. J. Poundstone. *Journal of the American Medical Association* 208: 360, 1969.

5. L. A. Young, L. G. Young, M. M. Klein, D. M. Klein, and D. Beyer. *Recreational Drugs.* Collier Books, New York, 1977, p. 52.

6. K. C. Osterhoudt, S. K. Lee, J. M. Callahan, and F. M. Henretig. *Veterinary & Human Toxicology* 39, 373-375, 1997.

7. G. R. Waller, G. H. Price, and E. D. Mitchell. *Science* 164: 1281-1282, 1969.

Cat's Claw

Cat's claw, as it is known in English, or uña de gato (Spanish for cat's claw), is represented by two species of the genus *Uncaria* of the family Rubiaceae indigenous to tropical South America. *Uncaria tomentosa* DC as well as *U. guianensis* (Aubl.) Gmel. are lianas or high-climbing, twining woody vines found in Amazonia. *U. guianensis* is collected in large quantities in South America for shipment to the European market. Most material supplied to the American market is designated *U. tomentosa.*

Among indigenous groups in the Amazon basin, a bark decoction of *U. guianensis* is used by the Piura as an anti-inflammatory, anti-rheumatic, and a contraceptive and for treating gastric ulcers and tumors. The Boras use the bark for the treatment of gonorrhea. In Colombia and Guiana, Indian groups use the stem bark for the treatment of dysentery. In South America, the plant has a reputation as a folk cancer remedy for cancers of the urinary tract, particularly in females. Uses of *U. tomentosa* from Peru, the center of the plant's range, mirror those of *U. guianensis.* It is also used for the treatment of gastric ulcers, arthritis, intestinal disorders, certain skin diseases, and various tumors.[1]

Much of the current excitement about cat's claw stems from the herb's treatment in Nicole Maxwell's *Witch Doctor's Apprentice: Hunting for Medicinal Plants in the Amazonian* (1990, Third Edition, Citadel Press, New York). She noted that the National Cancer Institute had reported encouraging results in anticancer screenings when funding for natural products research was cut in 1980. At the same time, use as a folk cancer remedy became popularized in South America, particularly in Brazil. Virtually unknown in the American herb market until the 1990s, cat's claw sales have evolved from a passing fad to an herbal ingredient available wherever herbal dietary supplements are sold.[2,3]

Biological activity of South American *Uncaria* species has been attributed to oxindole alkaloids. The major alkaloids in the leaves and stems of both species are isorhynchophylline and rynchophylline with their *N*-oxides, mitraphylline, dihydrocorynantheine, hirsutine, and hirsuteine. *U. tomentosa* also yields isomitraphylline and its *N*-oxide, along with dihydrocorynantheine *N*-oxide and hirsutine *N*-oxide.[4] Indole alkaloids from the roots of *U. tomentosa*, including isopteropodine, pteropodine, isomitraphylline, and isorynchophylline were found to have a pronounced immunostimulating effect in vitro and in vivo in the carbon clearance test.[5] Anti-inflammatory activity has been attributed to quinovic acid glycosides from the root of *U. tomentosa*, along with insignificant antiviral activity.[6] Antimutagenic and antioxidant (free radical scavenging) activity have been described for a root extract of *U. tomentosa* in vitro.[7] In a preliminary study, uncarine F, an alkaloid, has been shown to have a selective cytotoxic effect on leukemic cells, while not affecting normal stem cells in cell cultures.[8]

Recent research has shown that there are two chemical types of *Uncaria tomentosa* that differ greatly in their alkaloid content and consequently in their activity and potential therapeutic utility. One contains primarily the so-called pentacyclic (5-ring) oxindole alkaloids (pteropodine, isopteropodine, isomitraphylline, etc.). These compounds have immunostimulant properties. The other form contains primarily tetracyclic (4-ring) oxindole alkaloids (rhynchophylline and isorhynchophylline). Not only do these latter compounds act on the central nervous system, but they also antagonize any immunostimulant effect of the pentacyclic alkaloids.[9] Mixtures of these two types of cat's claw are therefore unsuitable for any therapeutic use unless certified to contain less than 0.02 percent tetracyclic oxindole alkaloids.[10]

The chemistry of cat's claw has been relatively well-studied over the past twenty years, along with pharmacological reports on specific fractions from the root and/or stem bark of both Amazonian species. The crude drug (consisting of both the bark of the root and the stem) is one of the most widely used and best known folk medicines, particularly in the upper Amazon basin, and is certainly worthy of closer scientific scrutiny.

Despite promising preliminary research on immunomodulating, anti-inflammatory, antimutagenic, and antitumor activity, controlled clinical studies are noticeably absent from the scientific literature.

However, there is no lack of conditions in which it is said to have been of clinical value, including Crohn's disease, diverticulitis, leaky bowel syndrome, colitis, hemorrhoids, gastritis, ulcers, and parasites, along with AIDS, arthritis, and cancer. One newsletter directed toward doctors (presumably including physicians) touted it as superior in healing benefit to the immune system to pau d'arco, echinacea, goldenseal, astragalus, Siberian ginseng, citrus seed extract, caprylic acid, shark cartilage, plus reishi and shiitake mushrooms.[11]

No significant toxicity has been reported from the widespread consumption of the herb or its preparations by American consumers, other than depleting consumers' bank accounts. Cat's claws' most dramatic effect has been in creating bulges in the wallets of its promoters. The hype that the herb could be of value in treatment of HIV infections, various forms of cancer, and dozens of other diseases is surpassed only by the hope of desperate patients seeking one last miracle cure when all else has failed.

REFERENCES

1. J. A. Duke and R. Vasquez. *Amazonian Ethnobotanical Dictionary.* CRC Press, Boca Raton, Florida, 1994, p. 172.

2. S. Foster. *Health Foods Business,* June 1995, pp. 24-25.

3. J. Jones. *Cat's Claw—Healing Vine of Peru.* Sylvan Press, Seattle, Washington, 1995, 152 pp.

4. S. R. Heminway and J. D. Phillipson. *Journal of Pharmacy and Pharmacology* 26 S: 113, 1974.

5. H. Wagner, B. Kreutzkamp, and K. Jurcic. *Planta Medica* 51: 419-423, 1985.

6. R. Aquino, V. De Feo, F. De Simone, C. Pizza, and G. Cirino. *Journal of Natural Products* 54(2): 453-459, 1991.

7. R. Rizzi, F. Re, A. Bianchi, V. De Feo, F. De Simone, L. Bianchi, and L. A. Stivala. *Journal of Ethnopharmacology* 38: 63-67, 1993.

8. H. Stuppner, S. Sturm, G. Geisen, U. Zillian, and G. Konwalinka. *Planta Medica* 59S: 583, 1993.

9. K.-H. Reinhard. *Zeitschrift für Phytotherapie* 18: 112-121, 1997.

10. G. Laus and K. Keplinger. *Zeitschrift für Phytotherapie* 18: 122-126, 1997.

11. P. N. Steinberg. *Townsend Letter* 130 (May): 1-2, 1994.

Celery Seed

Celery, *Apium graveolens* L., is an annual or biennial member of the family Apiaceae native to Eurasia, occurring in wild habitats in saline soils near coastal regions. Celery seed (actually a fruit) possesses a warm, aromatic, pungent flavor and is used as a condiment in food products. The essential oil distilled from the seeds is also used as a food flavoring.[1] Long cultivated as a garden vegetable in temperate climates, *A. graveolens* var. *dulce* (Mill.) Pers., with thickened, closely overlapping, ribbed petioles, is the familiar celery of grocery store produce sections. Another variety, *A. graveolens* var. *rapaceium* (Mill.), known as celeriac, has a greatly swollen taproot available as a specialty vegetable in the United States and as a staple in Europe.[2]

Celery seed has historically been reputed as a folk medicine for treating flatulence and indigestion, also as a diuretic, antispasmodic, and aphrodisiac. Primary uses included treatment of asthma, bronchitis, and rheumatism.[3]

As a major celery production region around the turn of the century, Kalamazoo, Michigan, was known as "celery city" and was home to the P. L. Abbey Company, once known as the Celery Medicine Company. The firm manufactured a number of celery-based patent medicines, including Kalamazoo Celery Nerve and Blood Tonic, Celerine compound, Celery Pepsin Bitters, and Kalamazoo Celery and Sarsaparilla Compound. The latter was touted as a cure for "biliousness, indigestion, dyspepsia, fever and ague, rheumatism, kidney and liver complaints, blood disorders, diseases of urinary organs, and all forms of nervousness, headache and neuralgia." It did not end there, for the preparation was also a "positive cure" for female complaints and disease arising from an impure state of the blood (translating to a cure for syphilis). In 1912, the company was prosecuted by the United States Department of Agriculture after the agency's Bureau of Chemistry determined that this and other P. L. Abbey Company products were mislabeled, as analysis of contents did not meet label statements. The company disappeared into oblivion by the 1920s, but it set the stage for the reentry of celery seed products sold today.[4]

The chemistry of celery seed and its essential oil, in particular, have been extensively studied. The essential oil contains delta-limonene, selinene, various sesquiterpene alcohols, and the characteristic flavor principles of celery, phthalides, primarily 3-*n*-butylphthalide, and sedanenolide. The seed also contains a fixed oil with ubiquitous fatty acids, including petroselinic, oleic, linoleic, myristic, palmitic, and stearic acids. Various pharmacological activities attributed to the seed or essential oil include antispasmodic, mild sedative, and anti-inflammatory activity, among others.[5]

Current interest in celery seed extracts stems from claims, as yet unproven, that it lowers uric acid levels and alleviates the pain of arthritis and rheumatism. It has also been suggested as a treatment for gout.[6] Until such claims are backed by peer-reviewed scientific research, celery seed extracts will hold a place in history next to celery patent medicines of the nineteenth century. The whole herb, root, and seed of celery are the subject of a German Commission E monograph. No therapeutic claims are allowed since effectiveness is not proven.[7]

Some individuals may suffer contact dermatitis or allergies from handling or ingesting celery. The psoralens in the stems or seeds can cause photodermatitis. Those allergic to birch pollen may experience hypersensitivity to celery ingestion (birch-celery syndrome). Despite traditional use of celery seed as a diuretic, the seed oil contains irritant compounds; therefore, ingestion of the seeds should be avoided in cases of genitourinary inflammation.[8]

REFERENCES

1. E. Guenther. *The Essential Oils,* Volume 4. D. Van Nostrand Co., New York, 1950, pp. 591-602.

2. M. Griffiths (Ed.). *Index of Garden Plants.* Timber Press, Portland, Oregon, 1994, p. 81.

3. J. E. Simon, A. F. Chadwick, and L. E. Craker. *Herbs: An Indexed Bibliography 1971-1980.* Archon Books, Hamden, Connecticut, 1984, pp. 19-20.

4. A. Palmeieri III. *Pharmacy in History* 39(3): 113-117, 1997.

5. A. Y. Leung and S. Foster. *Encyclopedia of Common Natural Ingredients Used in Food, Drugs, and Cosmetics*, Second Edition. John Wiley and Sons, New York, 1996, pp. 141-143.

6. J. A. Duke. *The Business of Herbs,* May/June: 12-13, 1997.

7. *Bundesanzeiger,* July 12, 1991.

8. H. Breiteneder, K. Hoffmann-Sommergruber, G. O'Riordain, M. Susani, H. Ahorn, C. Ebner, D. Kraft, and O. Scheiner. *European Journal of Biochemistry* 233(2): 484-489, 1995.

The Chamomiles and Yarrow

Three plants, closely related botanically and chemically, have long enjoyed great popularity as folk remedies useful in treating digestive disorders, cramps, various skin conditions, and minor infectious ailments. All three are members of the Asteraceae, or daisy family, and all yield a blue-colored volatile oil. Most volatile oils are local irritants, but those which contain quantities of the intensely blue azulene derivatives, together with other active ingredients, have anti-inflammatory properties.

Two of the plants are known as chamomile. The first consists of the flower heads of *Matricaria recutita* L., also referred to as *Chamomilla recutita* (L.) Rauschert or *Matricaria chamomilla* L.p.p. These three scientific names, all referring to the same plant, simply reflect taxonomic confusion. Commonly, the herb is known as German or Hungarian chamomile. The second chamomile consists of the flowerheads of *Chamaemelum nobile* (L.) All., also known, especially in the older literature, as *Anthemis nobilis* L. Its common names are Roman or English chamomile. The third plant in this general category consists of the flowering herb (entire overground plant) of *Achillea millefolium* L., commonly known as yarrow or milfoil. More recent taxonomic and phytochemical understanding of the *A. millefolium* group reveals that the widespread *A. millefolium* sensu stricto, abundant in Europe, is a hexaploid, while North American material once thought to be naturalized *A. millefolium* from Europe is a separate tetraploid taxon, now designated *A. lanulosa* Nutt. Chemical differences are associated with these genetic variations. The essential oil of hexaploid *A. millefolium* is azulene-free, while the essential oil of the tetraploid North American *A. lanulosa*, and of the European *A. collina* Becker, contains azulene and chamazulene. This knowledge can help explain why anti-inflammatory activity could not be associated with *A. millefolium* from Eastern Europe in a number of pharmacological assays.[1] Variation in azulene content has also been associated with soil type, with high amounts of proazulenes associated with yarrow growing conditions, correlating high levels of azulenes to soils with increased concentrations of available phosphate, magnesium, and manganese. Proazulene-free diploids and hexaploids are associated with nutrient-poor, acid, dry sites.[2]

Because of their popularity, all three plants are cultivated, but German chamomile is the principal article of commerce, both in the United States and on the European continent. Some 4,000 tons of chamomiles are produced annually throughout the world.[3] Their chemical constituents have been subject to detailed study. Interestingly, with the exception of their common component, the azulenes, the composition of the plants and their contained volatile oils is quite varied.[4]

Of the trio, German chamomile (the Germans refer to it as genuine chamomile) has been the most extensively investigated from the pharmacological and chemical viewpoints. It is used everywhere in Europe almost as a panacea, but basically as a carminative (aids digestion), an anti-inflammatory for various afflictions of the skin and mucous membranes, an antispasmodic primarily for treating menstrual cramps, and an anti-infective for all kinds of minor illnesses. Extracts of the plant or its volatile oil are used in the form of ointments, lotions, vapor baths, inhalations, and the like, all intended for local application. Internally, the drug is taken as a strong tea.[5]

Many believe that chamomile will cure almost anything; indeed, the Germans have a phrase for it, *alles zutraut,* meaning "capable of anything." As a popular remedy, it may be thought of as the European counterpart of ginseng. Some think that both patients and physicians would be better off if chamomile were even more widely used:

> How the Doctor's brow should smile
> Crown'd with wreaths of camomile.
>
> Thomas Moore
> *Wreaths for the Ministers*

German chamomile yields about 0.5 percent of a volatile oil, the principal anti-inflammatory and antispasmodic constituents of which are (−)-α-bisabolol, bisabololoxides A and B, and matricin.

In addition, certain flavonoids, such as apigenin and luteolin, are present in the flower heads and certainly contribute to both types of activity. The coumarins, herniarin and umbelliferone, also exhibit antispasmodic properties. Roman chamomile and yarrow contain similar, but not necessarily identical, active constituents.[6] It is clear that the therapeutic value of chamomile does not rest on a single constituent but on a complex mixture of chemically different compounds. In recent years, attention has focused on genetic and production factors in German chamomile that have identified improved, homogenous tetraploids, with stability of, and high yield in content of, (−)-α-bisabolol, chamazulene, *cis-trans*-en-in-dicycloether as well as the flavonoids apigenin and apigetrin.[7]

Since much of the value of the plant lies in its volatile oil, it is unfortunate that even a strong tea, properly prepared in a covered vessel and steeped for a long time, contains only about 10 to 15 percent of the volatile oil originally present in the plant material.[8] Whole extracts of the drug or preparations containing quantities of the volatile oil are certainly more effective and are increasingly marketed in the United States. Nevertheless, Farnsworth and Morgan believe that when the tea is used over a long period, beneficial effects may accumulate.[9] This belief is supported by the longtime usage of chamomile as a valued folk remedy in Europe and by its increasing popularity in the United States.

However, we must conclude these generally favorable comments on the chamomiles and yarrow with a word of caution. Because all three drugs contain varying amounts of allergens as well as pollen, tea made from any of them may cause contact dermatitis, anaphylaxis, or other hypersensitivity reactions in allergic individuals.[10] These reactions are, however, relatively infrequent. A survey of the worldwide literature revealed only about fifty reports of allergies resulting from the use of chamomile between the years 1887 and 1982.[11] Of these, only five were attributed to *Matricaria recutita* (German chamomile). Most of the others were caused by *Anthemis* species. The relative infrequency of chamomile hypersensitivity should certainly not deter normal persons from consuming it, if they so desire. However, persons known to be allergic to ragweed, asters, chrysanthemums, or other members of the family Asteraceae should be cautious about drinking tea prepared from the chamomiles or yarrow.

REFERENCES

1. R. R. Chandler, S. N. Hooper, and M. J. Harvey. *Economic Botany* 36: 203-223, 1982.

2. A. Preitchopf, B. Michler, and C.-G. Arnold. *Planta Medica* 55: 596-597, 1989.

3. C. Mann and E. J. Staba. The Chemistry, Pharmacology, and Commercial Formulations of Chamomile. In *Herbs, Spices, and Medicinal Plants: Recent Advances in Botany, Horticulture, and Pharmacology,* Volume 1, L. E. Cracker and J. E. Simon (Eds.). Oryx Press, Phoenix, Arizona, 1986, pp. 235-280.

4. R. C. Wren: *Potter's New Cyclopaedia of Botanical Drugs and Preparations,* Revised Edition. C. W. Daniel Company Limited, Saffron Walden, England, 1988, pp. 70-71, 290.

5. H. Schilcher. *Die Kamille.* Wissenschaftliche Velagsgesellschaft mbH, Stuttgart, Germany, 1987, 152 pp.

6. E. Steinegger and R. Hänsel. *Lehrbuch der Pharmakognosie und Phytopharmazie,* Fourth Edition. Springer-Verlag, Berlin, 1988, pp. 293-295, 308-313, 317-319.

7. W. Letchamo, R. Marquard, and W. Friedt. *Journal of Herbs, Spices & Medicinal Plants* 2: 19-26, 1994.

8. V. E. Tyler, L. R. Brady, and J. E. Robbers. *Pharmacognosy,* Ninth Edition. Lea and Febiger, Philadelphia, Pennsylvania, 1988, pp. 466-467.

9. N. R. Farnsworth and B. M. Morgan. *Journal of the American Medical Association* 221: 410, 1972.

10. M. Abramowicz (Ed.). *Medical Letter on Drugs and Therapeutics* 21(7): 30, 1979.

11. B. M. Hausen, E. Busker, and R. Carle. *Planta Medica* 50: 229-234, 1984.

Chaparral

Chaparral refers broadly to any dense thicket of shrubs or dwarf trees. More specifically, in recent herbal literature, it designates the leaflets of *Larrea tridentata* (Sessé & Moc.) Coville, a name considered by modern authors to be synonymous with *L. divaricata* Cav. and *L. mexicana* Moric. This strong-scented, olive green bush of the family Zygophyllaceae is the dominant shrub in the desert regions of the southwestern United States and Mexico. Better-known common names of the plant are creosote bush and greasewood.

An aqueous extract of the leaves and twigs, so-called chaparral tea, is an old Indian remedy and has been used for a wide variety of ailments, including arthritis, cancer, venereal disease, tuberculosis, bowel cramps, rheumatism, and colds. It is said to possess analgesic, expectorant, emetic, diuretic, and anti-inflammatory properties. One of its more unusual applications is that of a hair tonic.[1] Another is its purported property of "taking the residue of LSD out of the system . . . so you will have no recurrences of hallucinations."[2]

As might be expected, most of the attention focused on chaparral tea in recent years has concerned its use, and that of its principal ingredient, nordihydroguaiaretic acid (NDGA), as an anticancer agent. NDGA is a potent antioxidant, especially for fats and oils. As such, it was once thought to be potentially useful in the treatment of cancer. Early studies in rats did indicate that NDGA exerted an inhibitory effect on some tumor cells, but follow-up studies with the tea involving human beings have to date proved equivocal.[3] Besides, NDGA possesses considerable toxicity; long-term feeding studies in rats induced lesions in the mesenteric lymph nodes and kidneys. As a result, the compound was removed from the FDA's "Generally Recognized as Safe" (GRAS) list in 1968. However, it must be noted that the U.S. Department of Agriculture, which controls the use of antioxidants in lard and animal shortenings, still permits NDGA to be added to them.[4]

Recently, the anticancer activity of the South American subspecies was explored in a study on the effect of a leaf extraction on mammary gland cancer in female rats. Twenty days after the appearance of artificially induced tumors, the extract was given subcutaneously in a dose of 25mg/kg. Seventy-five tumors were treated with the extract, resulting in a reduction in 13 percent, a stabilization of tumors in 80 percent, and an increase in only 6 percent of the tumors. By comparison, none of the eighty tumors in the control group decreased, and only 15 percent remained stable. The survival time of the treated animals was significantly higher than the control animals. These results have not been duplicated in humans.[5]

In 1990, details of a case of liver disease resulting from the consumption of chaparral tablets were reported in the medical literature. The patient, a thirty-three-year-old woman, had been taking up to fifteen tablets daily for about five months in the belief that it would bring about improvement in a benign breast lump. She was subsequently diagnosed as suffering from subacute hepatic necrosis. Attending physicians ascribed the condition to drug-related toxicity, which eventually regressed with supportive care and abstinence from chaparral.[6] Two more cases were reported by the Centers for Disease Control on October 30, 1991. A third case was reported in November by a Wisconsin physician, and in early December 1991, a fourth case was diagnosed at the University of Chicago Hospital, which proved to be the most serious, involving a comatose woman with liver and kidney failure, who later recovered. On December 10, 1992 the FDA Center for Food Safety and Applied Nutrition issued a press release warning of the potential link between use of the herb and liver toxicity. Consequently, chaparral products were removed from store shelves, although they have quietly reappeared since then. From this it must be concluded that the FDA is indeed correct in considering chaparral unsafe for human consumption. Besides, the herb has no proven medical value.

Mosquitoes, however, have come to view chaparral and its contained NDGA quite differently. When fed the latter compound, the mosquito *Aedes aegypti* lengthened its average life span from twenty-nine to forty-five days.[7] Would it not be nice if NDGA had the same effect on human beings? Researchers, get busy!

REFERENCES

1. B. N. Timmerman. In *New World Deserts,* T. J. Mabry, J. H. Hunziker, and D. R. DiFeo Jr. (Eds.). Dowden, Hutchinson and Ross, Stroudsburg, Pennsylvania, 1977, pp. 252-256.

2. N. Baird. *Herbalist* 3(6): 6-8, 1978.

3. D. Mowrey. *Herbalist* 3(6): 28-30, 1978.

4. R. Winter. *A Consumer's Dictionary of Food Additives,* Revised Edition. Crown Publishers, Inc., New York, 1984, p. 176.

5. C. Anesini, J. Boccio, G. Cremaschi, A. Genaro, M. Zubillaga, L. B. Sterin, and E. S. Borda. *Phytotherapy Research* 11: 521-523, 1997.

6. M. Katz and F. Saibil. *Journal of Clinical Gastroenterology* 12: 203-206, 1990.

7. S. Boxer (Ed.). *Discover* 8(1): 13-14, 1987.

Chickweed

Despite the fact that it is prominently listed in almost every catalog of herbs currently available, and also that many writers describe it as a valuable herb (Gibbons,[1] for example, devotes six pages to the "useful" chickweed), I can think of no good reason to allow space to this worthless weed. Nevertheless, because people are apt to be misled by uncritical advocates of the medicinal use of chickweed, it is probably worthwhile to give it brief consideration.

Chickweed consists of the leaves and stems of *Stellaria media* (L.) Vill. of the family Caryophyllaceae, a low-growing or procumbent annual herb with small, white, star-shaped flowers. It is found throughout the world, having become a serious weed in many areas. Modern herbal advocates recommend chickweed for a large number of maladies ranging from constipation to hydrophobia.[2] A poultice of the plant is to be applied locally for every type of skin disease, including boils, abscesses, and ulcers. Taken internally, chickweed is supposed to be useful in curing bronchial asthma, stomach and bowel problems, blood disorders, lung disease, and obesity.

The plant is edible and makes tasty salads or cooked greens. As with many other green leafy vegetables, it contains some vitamin C (0.1 to 0.15 percent), accounting for its reputation as a cure for scurvy. Other constituents include the flavonoid glycoside rutin and various plant acids, esters, and alcohols.[3] Although there is an extensive scientific literature devoted to chickweed, there is no indication in it that any of the plant's constituents possess pronounced therapeutic value; indeed, most writings concern various methods of controlling this pesky weed.

Let us not waste any more time and space on the imagined medicinal value of this ineffective herb.

REFERENCES

1. E. Gibbons. *Stalking the Healthful Herbs,* Field Guide Edition. David McKay, New York, 1970, pp. 175-179.

2. W. Smith. *Wonders in Weeds.* Health Science Press, Bradford, England, 1977, pp. 48-50.

3. P. H. List and L. Hörhammer (Eds.). *Hagers Handbuch der Pharmazeutischen Praxis,* Fourth Edition, Volume 6B. Springer-Verlag, Berlin, 1979, pp. 526-527.

Chicory

Like many Americans, I first tasted chicory in New Orleans. The café au lait served on the patio of the Café du Monde Coffee Stand overlooking Jackson Square and the Mississippi River levee is a mixture of strong chicory coffee and hot milk, about half and half. Accompanied by crisp, hot beignets dusted with powdered sugar, the beverage provides a delightful experience. To my taste, it was bitter but mellow, two terms I previously thought contradictory!

Chicory, or succory, known botanically as *Cichorium intybus* L., is a perennial member of the daisy family (Asteraceae), native to Europe but now found growing wild along roadsides and in neglected fields throughout North America. Attaining a height of three to five feet or more, it is conspicuous for its attractive azure blue flowers. The plant has been grown in large quantities in Europe for many years to supply the demands of the beverage industry for roasted chicory root as a coffee additive or substitute. There is also a demand for the leaves, which are used in salads and as greens. As a consequence, many cultivated varieties exist that differ primarily in the size and texture of their roots and leaves.[1]

In folk medicine, chicory root is valued primarily as a mild nonirritating tonic with associated diuretic and, particularly, laxative effects.[2] It is said to protect the liver from, and act as a counterstimulant to, the effects of excessive coffee drinking.[3] Chicory root is valued in Egypt as a folk remedy for tachycardia (rapid heartbeat). The bruised leaves are considered a good poultice and are applied locally for the relief of swellings and inflammations. In addition, they are valued as a leafy green vegetable.

A rather large number of chemical constituents have been identified in chicory root, but none is especially physiologically active. They include 11 to 15 percent (up to 58 percent in cultivated plants) of the polysaccharide inulin, 10 to 22 percent of fructose, the bitter principles lactucin and lactucopicrin, tannin, both a fatty and a volatile oil, and small amounts of several other compounds.[4] From the culinary viewpoint, the inulin is particularly interesting. Upon roasting, it is converted to oxymethylfurfurol, a compound with a coffeelike aroma.

For more than fifty years, a scientific report has existed in the literature that lactucin and, to a lesser extent, lactucopicrin produce a sedative effect on the central nervous system and are capable of antagonizing the stimulant properties of caffeine beverages.[5] This work, carried out in rabbits and mice, may explain some of the old tales about chicory countering the undesirable "nervous" effects of coffee. However, much more study is needed, including quantitative measurements of the bitter principles in various varieties of the root, before a definite conclusion can be reached.

An investigation in the 1970s by Egyptian scientists studying the folkloric reputation of chicory as a drug useful in treating tachycardia apparently showed the presence of a digitalis-like principle in both the dried and roasted root that decreased the rate and amplitude of the heartbeat.[6] Its effects were demonstrated in the toad heart, and the activity in different samples, which incidentally varied greatly, was measured by the Baljet reaction, a color test for cardioactive glycosides. The active principle was not isolated, and the significance of this finding is very difficult to assess without additional information.

In Pakistan, the root has been used as a folk medicine for liver disease. Recently, researchers isolated a phenolic compound, esculetin, from the roots and confirmed hepatoprotectant activity in mice against paracetamol and carbon tetrachloride-induced hepatic damage.[7] Hepatoprotective activity has been linked to the ability of an aqueous extract of the root to inhibit oxidative degradation of DNA in tissue debris.[8]

Since chicory has been consumed in such large quantities by so many people for so many years without any reported untoward effects except an occasional allergy,[9] it is difficult to believe that it has the ability to produce any pronounced physiological or therapeutic actions in human beings. The conclusion is inevitable. Chicory is certainly as safe and has much less effect on the nervous system and the heart than the caffeine-rich coffee with which it is usually mixed. I think it is still all right to patronize the Café du Monde.

REFERENCES

1. M. Stuart (Ed). *The Encyclopedia of Herbs and Herbalism.* Grosset and Dunlap, New York, 1979, p. 173.

2. R. C. Wren. *Potter's New Cyclopaedia of Botanical Drugs and Preparations,* Revised Edition. The C. W. Daniel Company Limited, Saffron Walden, England, 1988, p. 74.

3. M. Grieve. *A Modern Herbal,* Volume 1. Dover Publications, New York, 1971, pp. 197-199.

4. P. H. List and L. Hörhammer (Eds.). *Hagers Handbuch der Pharmazeutischen Praxis,* Fourth Edition, Volume 4. Springer-Verlag, Berlin, Germany, 1973, pp. 3-5.

5. A. W. Forst. *Naunyn-Schmiedebergs Archiv für experimentelle Pathologie und Pharmakologie* 195: 1-25, 1940.

6. S. I. Balboa, A. Y. Zaki, S. M. Abdel-Wahab, E. S. M. El-Denshary, and M. Motazz-Bellah. *Planta Medica* 24: 133-144, 1973.

7. A. H. Gilani, K. H. Janbaz, and B. H. Shah. *Pharmacology Research* 37: 31-35, 1998.

8. S. Sultana, S. Perwaiz, M. Iqbal, and M. Athar. *Journal of Ethnopharmacology* 45: 189-192, 1995.

9. M. C. R. Symons. *Lancet* II: 1027, 1988.

Coltsfoot

Coltsfoot, the dried leaves and/or flower heads of *Tussilago farfara* L., is one of those plants whose botanical name reflects its medicinal application. *Tussilago* derives from the Latin *tussis,* meaning cough, and the plant has long been used to treat that affliction. This member of the family Asteraceae is a low, perennial, woolly herb that early in the spring produces a flowering stem with a single terminal yellow flower head. After the flower stem dies down, the hoof-shaped leaves appear. The plant is native to Europe but grows widely in moist, sandy places in the northeastern and north central United States and southern Canada.[1] Since the flowers and leaves appear at different times, they are collected and marketed separately.

Over the years, coltsfoot has been a very popular folk remedy for coughs and bronchial congestion. The leaves, the blossoms, and even the roots are ingredients in a large number of proprietary tea mixtures that are marketed in Europe for treating these conditions.[2] Gibbons has given recipes for preparing coltsfoot cough drops, coltsfoot cough syrup, and coltsfoot tea.[3] He also provided a formula for an herbal smoking mixture, similar to British Herb Tobacco, that consists principally of coltsfoot and is smoked to "cure coughs and wheezes." Since the principal active ingredient in the plant is a throat-soothing mucilage, smoking coltsfoot is certainly not rational therapy. The mucilage would be destroyed by burning, and the effect of smoke on already irritated mucous membranes would be increased irritation. Inhaling the vapors from coltsfoot leaves placed in a pan of simmering water, as suggested by Bricklin,[4] is again without value. The useful mucilage is not volatile and would not reach the affected tissues.

A scientific study carried out in Japan revealed some rather disturbing information about coltsfoot. The investigators analyzed dried young flowers because they are the parts widely used as an herbal remedy in Japan. They found the hepatotoxic (poisonous to the liver) pyrrolizidine alkaloid senkirkine to be present in relatively small amounts (0.015 percent). When rats were fed diets containing various amounts of coltsfoot, those which received high concentrations (greater than 4 percent) developed cancerous tumors of the liver. The scientists concluded that "it is evident that the young, pre-blooming flowers of coltsfoot are carcinogenic, showing a high incidence of hemangioendothelial sarcoma of the liver (8/12, 66.6 percent)."[5]

These data highlight the toxicity of pyrrolizidine alkaloids with an unsaturated pyrrolizidine nucleus, even at extremely low levels.[6] Two such alkaloids in the leaves and flowers, senkirkine and senecionine, are easily extracted in hot water. Continued or prolonged exposure to these pyrrolizidine alkaloids may have a cumulative effect.

For some time it was hoped that the rest of the plant might be devoid of pyrrolizidine alkaloids. However, a subsequent investigation of coltsfoot leaves showed senkirkine to be present in them as well.[7]

People suffering from throat irritations can no longer consider coltsfoot preparations appropriate therapy. Neither the flowers, the leaves, nor the roots can safely be used for medicinal purposes.[8] If readers want an herbal demulcent (soothing agent), they should consider a drug such as slippery elm bark or marshmallow root, both of which have long held official status in *The United States Pharmacopeia* (USP) and *The National Formulary* (NF).[9]

REFERENCES

1. H. W. Youngken. *Textbook of Pharmacognosy,* Sixth Edition. The Blakiston Company, Philadelphia, Pennsylvania, 1948, pp. 888-889.

2. P. H. List and L. Hörhammer (Eds.). *Hagers Handbuch der Pharmazeutischen Praxis,* Fourth Edition, Volume 6C. Springer-Verlag, Berlin, 1979, pp. 324-329.

3. E. Gibbons. *Stalking the Healthful Herbs,* Field Guide Edition. David McKay Company, New York, 1970, pp. 29-33.

4. M. Bricklin. *The Practical Encyclopedia of Natural Healing*. Rodale Press, Emmaus, Pennsylvania, 1976, p. 248.

5. I. Hirono, H. Mori, and C. C. J. Culvenor. *Gann* 67: 125-129, 1976.

6. C. A. Newall, L. A. Anderson, and J.D. Phillipson. *Herbal Medicines: A Guide for Health-Care Professionals.* The Pharmaceutical Press, London, 1996, pp. 85-86.

7. L. W. Smith and C. C. J. Culvenor. *Journal of Natural Products (Lloydia)* 44: 129-152, 1981.

8. *Bundesanzeiger,* July 27, 1990.

9. E. P. Claus and V. E. Tyler Jr. *Pharmacognosy,* Fifth Edition. Lea and Febiger, Philadelphia, Pennsylvania, 1965, p. 85.

Comfrey

Seldom does one encounter the degree of enthusiasm about anything that modern herbalists displayed in the 1970s and 1980s for common comfrey, the rhizome and roots as well as the leaves of *Symphytum officinale* L. (family Boraginaceae). Since Russian comfrey, derived from *Symphytum* x *uplandicum* Nym. (a hybrid between *S. officinale* and *S. asperum* Lepech, known as prickly comfrey), has become a very popular source of the commercial herb, the definition of comfrey must be extended to include it. Various writers emphasize comfrey's nearly universal healing properties as well as its safety in such statements as the following: "the first of all the 'wonder drugs' ";[1] "One of the most important therapeutic agents ever discovered by man";[2] "a most unusual plant with many preventive and curative properties";[3] "has a healing and soothing effect upon every organ it contracts";[4] "an ideal herb for making home remedies for use by amateur herbalists . . . nonpoisonous and completely harmless";[5] "a safe and harmless remedy";[6] "Toxicity is unlikely even after ingestion of moderately large quantities. . . ."[7] The list of quotations could go on and on, but the hyperbole remains the same.

Basically, comfrey is used in folk medicine in the form of an externally applied poultice for healing wounds. It is also taken internally as a tea or blended plant extract (so-called green drink) to heal stomach ulcers and to act as a "blood purifier."[8] Less-restrained advocates preach its virtue in treating cuts and wounds, burns, respiratory ailments of the lungs and bronchial passages, and ulcers of the bowels, stomach, liver, and gallbladder.[9] It is even said to facilitate the healing of broken bones, but this almost certainly comes from a misunderstanding of one of the common names of the plant, knitbone. It may have once been used to reduce the swelling and inflammation around a broken bone, but not to heal the bone itself.

Whatever healing properties comfrey may have are probably caused by its content of allantoin, an agent that promotes cell proliferation. Quantities of tannin and mucilage are also present. The underground parts contain 0.6 to 0.7 percent allantoin and 4 to 6.5 percent tannin; the leaves are poorer in allantoin, containing only about 0.3 percent, but richer in tannin, 8 to 9 percent. Large amounts of mucilage are present in both roots and leaves.[10] Much has been made of the vitamin B_{12} content of comfrey, but compared to the more customary natural source, liver, the concentration in the plant is not especially high.[11]

Although comfrey has been one of the most common herbs sold to the American public over the past thirty years, there is reason to believe that using it internally is definitely hazardous to the health. All comfrey species investigated have been found to contain hepatotoxic pyrrolizidine alkaloids (PAs), but the literature on the subject is confused due to a glaring lack of attention to proper botanical identification of the various *Symphytum* species studied. Common comfrey contains principally 7-acetylintermedine and 7-acetyllycopsamine in addition to their unacetylated precursors and symphytine. It does not contain high levels of echimidine, probably the most toxic comfrey PA. Echimidine has been identified, along with symphytine and six other PAs, in Russian comfrey. Russian comfrey was heavily promoted in the back-to-the-land movement, and most of the comfrey cultivated in home gardens is the Russian hybrid. The former alkaloid is also present in prickly comfrey. Comfrey root contains about ten times the concentration of PAs found in the leaves.[12,13]

Using echimidine as a marker, it becomes relatively easy to determine if samples of common comfrey are properly labeled. A Canadian study of thirteen commercial samples labeled either "comfrey" or "comfrey/*Symphytum officinale*" revealed that six of them contained echimidine and were therefore probably not derived from common comfrey (*S. officinale*) but from prickly comfrey (*S. asperum*) or Russian comfrey (*S.* x *uplandicum*).[14] Products containing echimidine are barred from sale in Canada for medicinal purposes, but, lacking chemical analysis, this cannot be determined accurately because the commercial labeling is so unreliable. However, the specific prohibition against echimidine products should not cause anyone to underestimate the potential danger of common comfrey that does contain other hepatotoxic PAs. All comfrey root–containing products are no longer acceptable in Canada.

Concerns about comfrey toxicity have apparently been justified by several reports in the clinical literature. In 1985, chronic PA intoxication was observed in a forty-nine-year-old woman who consumed both comfrey tea and comfrey-pepsin pills over a four-month period. She displayed the typical symptoms of hepatic veno-occlusive disease, a condition that may lead to liver failure.[15] Similar symptoms were reported in 1987 in a thirteen-year-old boy who had been treated for inflammatory bowel disease with comfrey-containing herbal tea over a two- to three-year period.[16] The same condition was observed in 1990 in a twenty-three-year-old man who had eaten young comfrey leaves as a vegetable. In this case, the subject died from liver failure.[17] It has been concluded that the most common cause of veno-occlusive disease is ingestion of plant materials that contain hepatotoxic PAs.[18]

Despite the well-documented toxicity of comfrey, the herb continues to be utilized to a limited extent, a situation probably reflective of its once enormous popularity. The German health authorities continue to allow the sale of both the leaves and the even more toxic root, provided daily consumption does not exceed 10 μg for the former and 1 μg for the latter of PAs with 1,2 unsaturated necine moieties inclusive of their *N*-oxides. Use must not exceed six weeks per year.[19] Since no comfrey on the American market is standardized to permit such dosage levels, these strictures are meaningless here.

Although the PA toxicity is the most significant hazard associated with the use of comfrey, it is not the only one. Several cases of atropine poisoning following the consumption of comfrey products have appeared in the literature.[20] The plant that is the most likely cause of such contamination is deadly nightshade (*Atropa belladonna* L.), the leaves of which could be mistaken for comfrey by inexperienced collectors. Foxglove (*Digitalis purpurea* L.), a source of cardiotonic glycosides, is a biennial that sends up its display of showy flowers in the second year. In the first year of growth, the basal rosettes of leaves to the untrained eye superficially resemble comfrey leaves. Several cases of digitalis poisoning have resulted from mistaking foxglove leaves for comfrey leaves.

Such occurrences emphasize the relative lack of quality control in some segments of the herb industry.

Regulatory agencies have taken at least limited action on the availability of comfrey in Canada, England, and Germany. The FDA has not taken regulatory action against this potentially dangerous and popular herb. However, the agency has proposed comfrey, goldenseal, and saw palmetto for in vivo carcinogenicity and reproductive toxicity testing on behalf of the National Toxicology Program Interagency Committee for Chemical Evaluation and Coordination. There is no need to spend federal money on investigating comfrey's toxicity. It is clearly established in the existing scientific literature.

One comprehensive book on comfrey bears a label on its back cover reminiscent of the required health-hazard warning on cigarette packages. This one reads, in part, ". . . we say that you or your animals should not eat, drink, or take comfrey raw, cooked, flour, tablets, or tea. There is a risk in internal use."[21] This statement says it all. It is most unfortunate that this poisonous herb continues to be sold to uninformed consumers by organizations that are ostensibly interested in promoting the good health of their customers.

REFERENCES

1. L. D. Hills. *Comfrey.* Universe Books, New York, 1976.
2. J. R. Christopher. *Herbalist* 1(5): 161-166, 1976.
3. G. J. Binding. *About Comfrey.* Thorsons Publishers Ltd., Wellingborough, England, 1974, p. 23.
4. M. Tierra. *The Way of Herbs.* Unity Press, Santa Cruz, California, 1980, p. 89.
5. E. Gibbons. *Stalking the Useful Herbs,* Field Guide Edition. David McKay Company, New York, 1970, p. 58.
6. W. Smith. *Wonders in Weeds.* Health Science Press, Bradford, England, 1977, p. 57.
7. D. G. Spoerke Jr. *Herbal Medications.* Woodbridge Press Publishing Co., Santa Barbara, California, 1980, p. 62.
8. T. Messina. *Herbalist New Health* 6(3): 28-29, 1981.
9. J. R. Christopher, op. cit., 1(5): pp. 161-166, 1976.
10. P. H. List and L. Hörhammer (Eds.). *Hagers Handbuch der Pharmazeutischen Praxis,* Fourth Edition, Volume 5B. Springer-Verlag, Berlin, 1979, pp. 706-710.
11. L. D. Hills, op. cit.
12. D. V. C. Awang. *Canadian Pharmaceutical Journal* 120: 100-104, 1987.
13. D. V. C. Awang. *HerbalGram* 25: 20-23, 1991.

14. R. J. Huxtable and D. V. C. Awang *American Journal of Medicine* 89: 547-548, 1990.

15. P. M. Ridker, S. Ohkuma, W. V. McDermott, C. Trey, and R. J. Huxtable. *Gastroenterology* 88: 1050-1054, 1985.

16. C. F. M. Weston, B. T. Cooper, J. D. Davies, and D. F. Levine. *British Medical Journal* 295: 183, 1987.

17. M. L. Yeong, B. Swinburn, M. Kennedy, and G. Nicholson. *Journal of Gastroenterology and Hepatology* 5: 211-214, 1990.

18. P. M. Ridker and W. V. McDermott. *Lancet* II: 657-658, 1989.

19. *Bundesanzeiger,* June 17, 1992.

20. D. V. C. Awang and D. G. Kindack. *Lancet* II: 44, 1989.

21. L. D. Hills, op. cit.

Cranberry

The American cranberry, also known as the trailing swamp cranberry, is referred to scientifically as *Vaccinium macrocarpon* Ait. (family Ericaceae). Many varieties of the species are cultivated extensively in natural or artificial bogs throughout the United States, but especially in Massachusetts and Washington. The native American bog cranberry (*V. macrocarpon*) occurs in much of eastern North America from Newfoundland to Manitoba, south to Virginia, Ohio, and northern Illinois, and locally to the mountains of North Carolina. The designation "large-fruited" (macrocarpon) cranberry distinguishes it from the smaller-fruited European cranberry *V. oxycoccos* L., a diminutive plant that occurs throughout the cooler regions of the northern hemisphere, notably the north of Europe, but also in North America in bogs south to New Jersey, Pennsylvania, and Minnesota. The European cranberry has considerably smaller fruits, which turn brownish red and are not as appealing as *V. macrocarpon*.

Cranberry fruits have long been valued for their pleasant acidulous flavor and have become a mainstay in sauces, relishes, and jellies, which are especially favored at Thanksgiving time.

In 1923, American scientists showed that the urine of test subjects became more acidic after eating large amounts of cranberries.[1] Because bacteria favor an alkaline medium for growth, these investigators speculated that a diet which included cranberries might be helpful in preventing and treating recurrent urinary tract infections (UTI). This condition is especially prevalent among women and often causes considerable discomfort. Until the advent of sulfa drugs and antibiotics, conventional medical treatment was largely ineffective.

Interested in any potential cure, women suffering from UTI began to consume quantities of the commercially available cranberry juice cocktail and reported, anecdotally, very satisfactory results.

Recommendations of the treatment spread, not only by word of mouth, but in occasional articles in regional medical journals. One of the latter reported symptomatic relief from chronic kidney inflammation in female patients who drank six ounces of cranberry juice twice daily. However, in 1967, a study showed that consumption of the commercial cranberry juice cocktail did not acidify the urine sufficiently to produce an appreciable antibacterial effect.[2] Such negative findings did not appear to influence adversely consumption of the product by UTI sufferers who remained convinced of its effectiveness.

Subsequent studies have provided evidence that the effectiveness of cranberry juice is not due to its ability to acidify the urine but to an entirely different mechanism. It appears to inhibit the ability of the microorganisms to adhere to the epithelial cells that line the urinary tract, thus rendering the environment there less suitable for the growth of the bacteria that typically cause UTI.[3,4] The most common of these bacteria, *Escherichia coli,* produces two constituents known as adhesins that cause the organism to cling to the cells, where they multiply rapidly. This adhering ability is inhibited by two constituents of cranberry juice. One of the antiadhesin factors is fructose; the other is a polymeric compound of unknown nature. Of a number of fruit juices tested, only cranberry and blueberry (both species of the same genus, *Vaccinium)* contained this latter, high-molecular-weight inhibitor.[5] In addition to these antiadhesin factors, cranberries contain various carbohydrates and fiber, as well as a number of plant acids, including benzoic, citric, malic, and quinic.[6]

A randomized, double-blind, placebo-controlled study on 153 elderly women volunteers (average seventy-eight years of age) was published in 1994. During the six-month study, volunteers were randomly assigned to consume either 300 ml of cranberry juice cocktail per day or a placebo drink, made to look and taste like cranberry juice, but without cranberry content. The study was designed to measure whether cranberry juice has an effect on bacteriuria (the passage of bacteria in the urine) or pyuria (presence of pus, indicating white blood cells, hence infection in the urine). At the end of the study, the researchers concluded that cranberry juice did reduce the frequency of both bacteriuria and pyuria in elderly women. This study provided the first good clinical evidence, in a relatively large sampling of patients, on the benefits of cranberry juice cocktail for UTI.[7]

Many persons consume three or more fluidounces daily of the cocktail (one-third of which is pure juice) as a preventive or twelve to thirty-two fluidounces daily as a treatment for UTI. Alternatively, capsules containing dried cranberry powder are now available; six capsules being equivalent to three fluidounces of cranberry juice cocktail. These contain more fiber and much less sugar than the cocktail, but an artificially sweetened cocktail is now available. It is also theoretically possible to consume either fresh or frozen cranberries: about 1.5 ounces equals three fluidounces of cocktail. However, in practice, this is not feasible because of their high acidity and extremely sour taste. An appropriate cranberry product does seem to be a useful adjunct in the prevention and treatment of urinary tract infections.

REFERENCES

1. N. R. Blatherwick and M. L. Long. *Journal of Biological Chemistry* 57: 815-818, 1923.

2. *Lawrence Review of Natural Products,* August, 1987.

3. A. E. Sobota. *Journal of Urology* 131: 1013-1016, 1984.

4. M. S. Soloway and R. A. Smith. *Journal of the American Medical Association* 260: 1465, 1988.

5. I. Ofek, J. Goldhar, D. Zafriri, H. Lis, and N. Sharon. *New England Journal of Medicine* 324: 1599, 1991.

6. B. G. Hughes and L. D. Lawson. *American Journal of Hospital Pharmacy* 46: 1129, 1989.

7. J. Avorn. *Journal of the American Medicinal Association* 271: 751-754, 1994.

Cucurbita

Seeds of several species of the genus *Cucurbita* have long enjoyed a considerable reputation as teniafuges (agents that paralyze and expel intestinal worms). Chief among these are pumpkin seeds or pepo, obtained from *C. pepo* L., but the seeds of the autumn squash (*C. maxima* Duchesne) and of the Canada pumpkin or crookneck squash [C. *moschata* (Duchesne) Poir.] have similar properties.[1] All are large edible fruits produced by herbaceous, running (vinelike) plants of the family Cucurbitaceae. Numerous cultivated varieties exist.

When used as a teniafuge or anthelmintic, cucurbita seeds are ordinarily administered in the form of the ground seeds themselves, as an infusion (tea), or as an emulsion made by beating the seeds with powdered sugar and milk or water. Usually three divided doses are given, representing a total weight of seeds ranging from 60 to as much as 500 grams. Such treatment is said to be effective in expelling both tapeworms and roundworms.[2] Another traditional use of the seeds is in the prevention and treatment of chronic prostatic hypertrophy (enlargement of the prostate gland) in males. A handful of the seeds eaten daily is supposed to be a very popular remedy for this condition in Bulgaria, Turkey, and the Ukraine.[3]

Cucurbitin, an unusual amino acid identified chemically as (−)-3-amino-3-carboxypyrrolidine, is the active principle responsible for the anthelmintic (worm-expelling) effects of the drug. It occurs only in the seeds of *Cucurbita* species, but its concentration is quite variable even in seeds of the same species. This variability probably accounts for reports in the literature that cucurbita seeds are either unreliable or ineffective as a teniafuge. One study showed the concentration of cucurbitin in different samples of *C. pepo* ranged from 1.66 to 6.63 percent, in *C. maxima* from 5.29 to 19.37 percent, and in *C. moschata* from 3.98 to 8.44 percent.[4]

Identifying the principle(s) responsible for any beneficial effects on the prostate gland is not so straightforward. The fatty oil contained in cucurbita seeds, in amounts approaching 50 percent, is an efficient diuretic,[2] so the increased urine flow it produces may give an illusory sense of reduction of prostatic swelling or hypertrophy. Administration of unsaturated fatty acids is thought by some to be beneficial in the treatment of prostate problems.[3] Cucurbita seed oil contains a number of these, including about 25 percent oleic acid and 55 percent linoleic acid.[5] Phytosterols may also play some role.[6] Delta-7 sterols contained in the seed have been shown to compete with dihydrotestosterone at androgen receptors for human fibroblasts.[7]

No satisfactory clinical evidence presently attests to the utility of either cucurbita seed or seed oil in treating prostatic enlargement. As a matter of fact, there is no more evidence to support such claims than for other deceptive claims made about the efficacy of a mixture of amino acids (glycine, alanine, and glutamic acid) in relieving the same condition. In 1990, the Food and Drug Administration banned the sale of all nonprescription drugs for the treatment of prostate enlargement because they might cause serious side effects and complications. Of course, the various products continue to be sold as "nutritional supplements," but without therapeutic claims.[8]

The German Commission E monograph on pumpkin seeds does allow products to be used in stages I and II of prostatic adenoma to relieve micturition difficulties at a daily dose of 10 g of the ground seeds or comparable preparations. Controlled clinical studies are needed to substantiate therapeutic claims.[7]

Cucurbita seeds are an effective teniafuge, but strain differences in cucurbitin content make it very difficult to know how much should be taken to get results. Although toxicity or undesirable side effects associated with cucurbita seeds have not been reported in the literature, it is not easy, since cucurbitin content varies so, to recommend the unstandardized drug for treatment of intestinal worms. Lacking supporting clinical evidence, it is impossible to recommend it for prostate problems.

REFERENCES

1. V. E. Tyler, L. R. Brady, and J. E. Robbers. *Pharmacognosy,* Ninth Edition. Lea and Febiger, Philadelphia, Pennsylvania, 1988, p. 469.

2. H. W. Felter and J. U. Lloyd. *King's American Dispensatory*, Eighteenth Edition, Volume 2. The Ohio Valley Co., Cincinnati, Ohio, 1900, pp. 1443-1444.
3. K. W. Donsbach. *Your Prostate.* International Institute of Natural Health Sciences, Huntington Beach, California, 1976, pp. 24.
4. V. H. Mihranian and C. I. Abou-Chaar. *Lloydia* 31: 23-29, 1968.
5. H. A. Hoppe. *Drogenkunde,* Eighth Edition, Volume 1. Walter de Gruyter, Berlin, 1975, pp. 368-369.
6. H. Schilcher, U. Dunzendorfer, and F. Ascali. *Urologe (Ausgabe B)* 27: 316-319, 1987.
7. V. Schulz, R. Hänsel, and V. E. Tyler. *Rational Phytotherapy: A Physician's Guide to Herbal Medicine,* Third Edition. Springer, Berlin, 1998, pp. 229-230.
8. Anon. *Health Foods Business* 36(9): 20, 1990.

Damiana

Damiana, from the leaves of the Mexican shrub *Turnera diffusa* Willd. var. *aphrodisiaca* (L.F. Ward) Urb., family Turneraceae, was introduced into American medicine in the fall of 1874 by a Washington, DC, druggist who sold eight-ounce bottles of the tincture for two dollars each.[1] The product was touted as a powerful aphrodisiac "to improve the sexual ability of the enfeebled and aged." It was said to have a specific effect on all the organs of the pelvis and to give "increased tone and activity to all the secretions in that vicinity." Stories of Mexican men who had sired children at very advanced ages were circulated to substantiate these claims.[2]

Within very few months, the reported activities of the drug were recognized as fraudulent,[3] but more than a century has scarcely diminished damiana's reputation in the minds of people who want to believe in its tonic and stimulating properties. Any physiological activity in various proprietary damiana preparations marketed around the turn of the century (e.g., Nyal's Compound Extract of Damiana) was actually due to the presence of other drugs, such as coca or nux vomica, and to a high alcoholic content, usually about 50 percent.[4]

Popular writers on drugs indicate that damiana leaves, drunk in the form of a tea or smoked like tobacco, produced a relaxed state in the user and a kind of subtle high with sexual overtones, somewhat reminiscent of the effects of marijuana.[5] The drug is supposed to be especially effective in women. Damiana liqueurs, produced in Mexico and subtly advertised as aphrodisiacs, contain only minute quantities of the drug. The amount is sufficient, however, to give these beverages a distinctive flavor.

Chemical studies have shown that damiana contains between 0.2 and 0.9 percent of a complex volatile oil which is responsible for most of the characteristic odor and taste of the drug. In addition, quantities of a resin, a bitter principle, tannin, mucilage, starch, etc., are present.[6] The reported presence of caffeine requires verifying. No constituent responsible for claims of damiana as an aphrodisiac has ever been identified. On the basis of available evidence, we must conclude that the drug lacks significant physiological activity and that no basis exists for its consumption by human beings. Indeed, the purported virtue of the product has been described as nothing more than an "herbal hoax."[7]

Writing in 1904, pharmacist John Uri Lloyd best captured its position in herbal medicine: "Damiana is a homely, domestic remedy, innocent of the attributes under which, in American medicine, it has, for a quarter of a century, been forced to masquerade."[8]

The masquerade has now continued for more than a century.

REFERENCES

1. Anon. *American Journal of Pharmacy* 47: 426-427, 1875.
2. J. M. Maisch. *American Journal of Pharmacy* 47: 380-381, 1875.
3. Ibid., pp. 429-430.
4. *Nostrums and Quackery,* Second Edition, Volume 1. American Medical Association, Chicago, Illinois, 1912, p. 537.
5. *High Times Encyclopedia of Recreational Drugs.* Stonehill Publishing Co., New York, 1978, pp. 106–107.
6. H. A. Hoppe. *Drogenkunde,* Eighth Edition, Volume 1. Walter de Gruyter, Berlin, 1975, p. 1096.
7. V. E. Tyler. *Pharmacy in History* 25: 55-60, 1983.
8. J. U. Lloyd. *The Pharmaceutical Review* 22: 126-131, 1904.

Dandelion

It is useful in treating warts, fungus infections, external and internal malignant growths, ulceration of the urinary passages, and obstructions of the liver, gallbladder, and spleen. It is a laxative, a stomach remedy, a promoter of healthy circulation, a skin toner, and a blood vessel cleanser and strengthener. It cures rheumatism, badly affected arthritic joints, and it is a marvelous tonic. It makes a fine wine, a great beer, an excellent coffee substitute, as well as a good food for birds, bees, pigs, rabbits, and people. [1]

What is it? The newest miracle drug or manna from heaven? No! It is the common dandelion. After reading this description of that plant's virtues by herbal advocate William Smith, I even felt guilty about mowing my lawn. Dandelion is one plant that probably requires no description. Its basal rosette of toothy leaves that rises from a hollow scape or flower stem capped by a deep yellow head of ligulate flowers makes *Taraxacum officinale* Weber (family Asteraceae) one of our best-known weeds. A native of Europe, it is ubiquitous in North America. What is not generally known is that there are literally hundreds of forms or subspecies of the plant that vary sufficiently in appearance from one another to be recognized by botanical specialists.[2]

A large number of constituents have been isolated from the underground parts (rhizome and roots) of dandelion, but it is difficult to attribute much therapeutic utility to any of them. An undefined bitter principle, designated taraxacin, apparently has some favorable influence on the digestive process, but the compound responsible for the mild laxative action of the drug is unknown.[3] In small animal experiments, extracts of the leaves have exhibited a pronounced diuretic action, but again, the responsible principle(s) remains unidentified.[4] A recent report found that a triterpene fraction of an ethanolic extract of a root had significant antiplatelet agreggation activity in human platelets.[5] Dandelion appears to be essentially free of significant toxicity or side effects. Some few individuals do develop a skin rash (allergic dermatitis) following repeated contact with the plant.[6]

In summary, no significant therapeutic benefits should be expected from the use of any dandelion products. The roots may stimulate the appetite a bit and aid digestion, as well as produce a slight laxative effect. Leaves of the plant may also have a transient diuretic action. However, many persons do enjoy dandelion greens, and they are a fairly good source of vitamin A. My personal experience with dandelion wine enhanced my appreciation of the California jug varieties without dandelions. Some say that the roasted root provides a tasty coffee substitute. These culinary applications far outweigh any medicinal uses for this common plant. I guess it is all right to mow my weedy lawn after all.

REFERENCES

1. W. Smith. *Wonders in Weeds.* Health Science Press, Bradford, England, 1977, pp. 66-68.
2. L. H. Bailey and E. Z. Bailey. *Hortus Third.* Macmillan, New York, 1976, p. 1097.
3. P. H. List and L. Hörhammer (Eds.). *Hagers Handbuch der Pharmazeutischen Praxis,* Fourth Edition, Volume 6C. Springer-Verlag, Berlin, 1979, pp. 16-21.
4. E. Ràcz-Kotilla, G. Ràcz, and A. Solomon. *Planta Medica* 26: 212-217, 1974.
5. H. Neef, F. Cilli, P. J. Declerck, and G. Laekeman. *Phytotherapy Research* 10: S138-S140, 1996.
6. *Lawrence Review of Natural Products,* December, 1987.

Devil's Claw

"What's this devil's claw good for?"
"Rheumatism. It's really good!"
"I never heard of it. If it's so good why don't they sell it in drugstores?"
"They can't. The FDA won't approve it. If they did, it would put all the doctors out of business."

Conversation in an Orlando,
Florida, health food store

Devil's claw consists of the secondary storage roots of *Harpagophytum procumbens* DC., a South African plant belonging to the family Pedaliaceae. The common name is derived from the plant's peculiar fruits, which seem to be covered with miniature grappling hooks. Devil's claw, the name commonly used in the United States, is actually a translation of the German *Teufelskralle;* English synonyms include wood spider and grapple plant.[1]

The drug has been recommended for treating a wide variety of conditions, including diseases of the liver, kidneys, and bladder, as well as allergies, arteriosclerosis, lumbago, gastrointestinal disturbances, menstrual difficulties, neuralgia, headache, climacteric (change of life) problems, heartburn, nicotine poisoning, and above all, rheumatism and arthritis.[2] The allegation that devil's claw induces abortion remains unverified.[3] It may be based on a misinterpretation of a statement by Watt and Breyer-Brandwijk that the drug is used by African natives to alleviate pain in pregnant women and especially in those anticipating a difficult delivery.[4] Even discounting this property, enough therapeutic activities were attributed to devil's claw to cause some to consider it a "wonder" drug.

A limited clinical study carried out in Germany reported, in 1976, that devil's claw exhibited anti-inflammatory activity comparable in many respects to the well-known antiarthritic drug, phenylbutazone.

Analgesic effects were also observed along with reductions in abnormally high cholesterol and uric-acid blood levels.[5] However, this is apparently the only study in animals or humans to demonstrate positive anti-inflammatory activity. Several investigators have tested the efficacy of devil's claw in various standard inflammation models in animals. Little or no activity was observed by any of them. This same lack of significant anti-inflammatory activity was demonstrated for harpagoside, the principal one of several iridoid glycosides occurring in devil's claw in a range of 0.1 to 3 percent.[6]

A recent randomized, double-blind clinical trial in Germany evaluated the effects of the drug on chronic low back pain in 118 patients over four weeks. Patients in the treatment group received the equvialent of 6 g of dried tuber per day, standardized to 50 mg harpagoside. Positive, though inconclusive, results in reducing or eliminating acute attacks of low back pain were reported, prompting a call for more clinical studies.[7]

Folkloric indications that devil's claw might prove useful as an antiarthritic and antirheumatic agent have not been verified in scientific studies. It continues to be widely utilized, particularly in Europe, as an appetite stimulant and a digestive aid. The authentic herb is, however, quite expensive, and many less costly drugs are equally effective for these purposes. Devil's claw apparently lacks any appreciable toxicity and is free from side effects. It is by no means a wonder drug for any condition.[8]

REFERENCES

1. O. H. Volk. *Deutsche Apotheker Zeitung* 104: 573-576, 1964.
2. R. Kämpf. *Schweizerische Apotheker Zeitung* 114: 337-342, 1976.
3. M. Abramowicz (Ed.). *Medical Letter on Drugs and Therapeutics* 21(7): 30, 1979.
4. J. M. Watt and M. G. Breyer-Brandwijk. *The Medicinal and Poisonous Plants of Southern and Eastern Africa,* Second Edition. E. & S. Livingstone Ltd., Edinburgh, 1962, p. 830.
5. R. Kämpf, op. cit., pp. 337-342.
6. F.-C. Czygan. *Zeitschrift für Phytotherapie* 8: 17-20, 1987.
7. S. Chrubasik, C. H. Zimpfer, U. Schütt and R. Ziegler. *Phytomedicine* 3: 1-10, 1996.
8. R. Jaspersen-Schib. *Deutsche Apotheker Zeitung* 130: 71-73, 1990.

Dong Quai

Dong quai, dang gui, or tang kuei consists of the root of the Chinese plant *Angelica polymorpha* Maxim. var *sinensis* Oliv., also known as *A. sinensis* (Oliv.) Diels, a member of the family Apiaceae. The drug is mildly laxative, although it is used primarily for its uterine tonic, antispasmodic, and alterative (blood purifying) effects.[1]

It is recommended by modern herbalists for the treatment of almost every gynecological ailment, including menstrual cramps, irregularity or retarded flow, and weakness during the menstrual period. Dong quai is also said to bring relief from the symptoms of menopause, but should not be used during pregnancy. In addition, it is thought to be a useful antispasmodic and of value in the treatment of hypertension. Its reputation also extends to blood "purification" and "nourishment," and finally, to treating constipation.[2]

Under chemical investigation, seven different coumarin derivatives have been identified in dong quai, including oxypeucedanin, osthole, imperatorin, psoralen, and bergapten.[3] Many coumarins are known to act as vasodilators and antispasmodics; others, such as osthole, have a stimulating action on the central nervous system.[4] Thus, at least some of the purported activities of dong quai could be accounted for by these compounds.

However, large doses of coumarins are not without undesirable effects, and the furocoumarins, such as psoralen and bergapten, are prone to cause photosensitization that may result in a type of dermatitis in persons exposed to them. In 1981, investigators concluded that these so-called psoralens present sufficient risks to humans that all unnecessary exposure to them should be avoided.[5] For this reason, large amounts of a furocoumarin-containing drug such as dong quai cannot be recommended. Substantial clinical evidence is lacking in Western scientific literature to support the effectiveness of dong quai for the various conditions for which it is advocated. The only U.S. study of the effects of dong quai on postmenopausal symptoms (night sweats, hot flashes) found it to be no more effective than a placebo.[6] Dong-quai is one of the most widely prescribed drugs in traditional Chinese medicine and is more widely used than other popular herbs such as ginseng.[7] It is generally used in combination with other ingredients. For acceptance in Western societies, controlled clinical studies should be conducted. Until then, there is little reason to utilize it as a therapeutic agent.

REFERENCES

1. The Revolutionary Health Committee of Hunan Province. *A Barefoot Doctor's Manual.* Running Press, Philadelphia, Pennsylvania, 1977, p. 721.

2. M. Tierra. *The Way of Herbs.* Unity Press, Santa Cruz, California, 1980, pp. 124-125.

3. K. Hata, M. Jozawa, and Y. Ikeshiro. *Yakugaku Zasshi* 87: 464-465, 1967.

4. E. Steinegger and R. Hänsel. *Lehrbuch der Pharmakognosie,* Third Edition. Springer-Verlag, Berlin, 1972, pp. 132-135.

5. G. W. Ivie, D. L. Holt, and M. C. Ivey. *Science* 213: 909-910, 1981.

6. S. Gilbert. *The New York Times,* January 13, 1998, p. B17.

7. S. Foster and C. X. Yue. *Herbal Emissaries: Bringing Chinese Herbs to the West.* Healing Arts Press, Rochester, Vermont, 1992, pp. 65-72.

Echinacea

Thoughtful readers may often wonder how humankind discovered the medicinal virtues of so many different plants. After teaching about drugs from natural sources for several years at the University of Nebraska, I knew the answer. Every few weeks during the summer months, another package would arrive in the mail containing some fragments of the woody rhizome and roots of a "peculiar" plant that, as explained in the accompanying letter, caused an unusual, acrid, tingling sensation on the tongue when it was chewed. Some Nebraska farmer, a close observer of nature, had once again discovered a potentially useful plant by the age-old process of trial and error.

I responded to all such inquiries with a much-used form letter identifying the plant source as *Echinacea angustifolia* DC., usually referred to simply as echinacea, or cone flower, or narrow-leaved or purple cone flower. Today, the more specific designation, narrow-leaved echinacea, is used to describe the plant. This perennial member of the daisy family (Asteraceae), with its narrow leaves and stout stem up to three feet in height, terminating in a single, large purplish flower head, is native to the central United States. Altogether, nine species of *Echinacea* grow in the United States. In addition to *E. angustifolia, Echinacea pallida* (Nutt.) Nutt., known as pale-flowered echinacea, and *Echinacea purpurea* (L.) Moench, the commonly cultivated garden variety, are all articles of commerce. Because of the tendency to employ cultivated plant material, *E. purpurea* is now the most commonly utilized species.[1]

Appropriately enough, echinacea was first introduced into medicine by a Nebraskan, Dr. H. C. F. Meyer of Pawnee City. Having learned of the therapeutic value of the drug from the Indians about 1871, Meyer used it to prepare a "blood purifier" that he claimed was useful in treating almost any condition, including rheumatism, migraine, erysipelas (streptococcus infections), dyspepsia, pain, wounds, sores, eczema, dizziness, sore eyes, poisoning by herbs, rattlesnake bites, tumors, syphilis, gangrene, typhoid, malaria, diphtheria, bee stings, hydrophobia, and hemorrhoids.[2] In 1885, Meyer called the attention of a pharmaceutical manufacturer, Lloyd Brothers of Cincinnati, to the drug, and that firm subsequently introduced several echinacea products intended primarily as anti-infective agents. By 1920, echinacea was the firm's most popular plant drug, but with the advent of the sulfa drugs in the 1930s, it began to fall into disuse. No longer valued in conventional medicine, the drug continues to be employed in folk medicine, either taken internally to increase the body's resistance to various types of infections or applied locally for its wound-healing action.[3]

Most of the scientific and clinical studies on echinacea have been carried out in Germany, primarily with dosage forms prepared from the fresh overground portion of *E. purpurea* that are intended to be administered by injection or applied locally. Injectable preparations, however, are not available in the United States. There remains considerable controversy as to the relative effectiveness of echinacea following oral administration. The herb is most readily available in this country in liquid form, specifically as a hydroalcoholic extract. Indeed, it has been suggested that such preparations are effective because echinacea stimulates lymphatic tissue in the mouth, thereby initiating an immune response. Assuming that to be the case, powdered echinacea administered orally in the form of capsules would probably be less active.[4]

Of the various activities attributed to echinacea, the one that is probably best substantiated is its immune-stimulant effect. This is said to be brought about by three different mechanisms: stimulating phagocytosis, increasing respiratory activity, and causing increased mobility of the leukocytes.[5] The exact identity of the principles responsible for this action remains unknown. Without question, high-molecular-weight polysaccharides are effective, but their stimulation of phagocytosis is apparently enhanced by components of the alkamide fraction (mainly isobutylamides), by glycoproteins, and by cichoric acid.[6,7]

The literature on echinacea has become so vast and the hyperbole of its advocates so extensive that it has become difficult to separate what we know about it with certainty from what may be true. Oral consumption of a hydroalcoholic extract of the fresh or recently dried whole plant (including the overground portion) seems to be especially useful in preventing and treating the common cold and conditions associated with it, such as sore throat. German authorities also recommend it as a supportive or auxiliary treatment for recurrent infections of the respiratory or urinary tracts.[8] Externally, it is useful in the treatment of hard-to-heal superficial wounds. Significant side effects have not been reported, but allergies are always possible, particularly with plants in this family.

At one time, echinacea was extensively adulterated with *Parthenium integrifolium* L., commonly known as prairie dock or Missouri snakeroot.[9] Even some of the early scientific studies were invalidated because echinacea was confused with this plant. Potential consumers of echinacea should make every effort to obtain the best quality product available. Careful investigation of the reputation of the manufacturer should precede the purchase of echinacea or, for that matter, any other plant extract.

In today's market, dominated by standardized extracts of the most popular herbs, most products containing *E. angustifolia* are standardized to a certain content of the caffeic acid glycoside echinacoside. This was the first compound to which biological activity was attributed for the genus. A 1950 study on the isolated compound found it had mild, insignificant antibacterial action. The compound was once considered a chemical marker for the identify of *E. angustifolia*; however, the compound has been identified in several additional echinacea species. Echinacoside has not been found to be involved in the herb's immunostimulatory activity. Therefore, it is meaningless as a chemical marker of identity or as a compound to predict biological activity. Why "standardize" an herb product to a meaningless compound?

Much more work on the efficacy of echinacea for various conditions in human subjects must be carried out before a definitive statement can be made regarding its utility as a modern therapeutic agent. However, it is a botanical that is deserving of continued attention by scientists and clinicians.

REFERENCES

1. S. Foster. *Echinacea: The Purple Coneflowers, Botanical Series No. 301,* Second Edition. American Botanical Council, Austin, Texas, 1996, 7 pp.

2. J. U. Lloyd. *A Treatise on Echinacea.* Lloyd Brothers, Cincinnati, Ohio, 1924.

3. R. Bauer and H. Wagner. *Zeitschrift für Phytotherapie* 9: 151-159, 1988.

4. C. Hobbs. *The Echinacea Handbook.* Eclectic Medical Publications, Portland, Oregon, 1989, 118 pp.

5. G. Harnischfeger. *Deutsche Apotheker Zeitung* 125: 1295-1296, 1985.

6. R. Bauer, P. Remiger, K. Jurcic, and H. Wagner. *Zeitschrift für Phytotherapie* 10: 43-48, 1989.

7. R. Bauer. Echinacea: Biological Effects and Active Principles. In *Phytomedicines of Europe Chemistry and Biological Activity.* L. D. Lawson and R. Bauer, (Eds.). American Chemical Society, Washington, DC, 1998, pp. 140-157.

8. *Bundesanzeiger,* January 5, 1989.

9. S. Foster. *Echinacea: Nature's Immune Enhancer.* Healing Arts Press, Rochester, Vermont, 1991, 150 pp.

Ephedra (Ma Huang)

Ephedra (ma huang), together with its principal alkaloid ephedrine, was perhaps the first of the Chinese herbal remedies to see significant use in Western medicine. Known in China for more than 5,000 years, the green stems of various *Ephedra* species, particularly *E. sinica* Stapf, *E. equisetina* Bunge, and others of the family Ephedraceae, were employed there, and *E. gerardiana* Wall. was used in India for the treatment of bronchial asthma and related conditions.[1]

The active constituent, ephedrine, was isolated by a Japanese chemist, N. Nagai, in 1887. However, it was not until 1924 when K. K. Chen and his mentor C. P. Schmidt, working at Peking Union Medical College, began to publish a series of papers on its pharmacological properties that physicians in this country began to appreciate the utility of the drug. Ephedrine became widely used as a nasal decongestant, a central nervous system stimulant, and a treatment for bronchial asthma.[2] Other alkaloids, pseudoephedrine, norephedrine, norpseudoephedrine, etc., with similar but not identical properties, were subsequently found in various *Ephedra* species.[3]

Studies on the herb revealed that the approximately forty species of *Ephedra* could be divided into several geographic types which seem to vary qualitatively and quantitatively in their alkaloid content. The significant finding regarding these types is that the North and Central American types all appear to be alkaloid free.[4] Thus, any activity attributed to these species must result from compounds other than ephedrine or its derivatives. For that reason, species such as *E. nevadensis* (see Mormon tea) are not considered to be in this group of ephedra plants and are discussed elsewhere. It should be noted that *Ephedra* species are often extremely difficult to distinguish from one another, even for the specialist.

Ephedra is a potent and useful herb for relieving the constriction and congestion associated with bronchial asthma. It is an effective nasal decongestant and is used in the treatment of various allergic disorders in adults. It acts as a strong central nervous system stimulant, but despite the claims of some advocates,[5] there is no substantial clinical evidence that it is either a safe or effective promoter of weight loss in obese persons or an enhancer of athletic performance.

Unfortunately, ephedra and its contained ephedrine also increase both systolic and diastolic blood pressure.[6] They also increase the heart rate and may cause palpitations as well as nervousness, headache, insomnia, and dizziness.[7] Although the herb may be a very useful one in the treatment of various asthmatic and congestive conditions, the side effects indicated render its indiscriminate use highly inadvisable, particularly in persons suffering from heart conditions, hypertension, diabetes, or thyroid disease.

Because ephedrine can serve as precursor for the illegal synthesis of methamphetamine or "speed," a common drug of abuse, several states have passed laws regulating the sale of the alkaloid or products containing it. Although various species of ephedra contain 0.5 to 2.5 percent of an alkaloid mixture, some 30 to 90 percent of which is ephedrine, it must be emphasized that the herb is no longer the principal source of commercial ephedrine. That compound is produced today by chemical synthesis involving the reductive condensation of L-l-phenyl-l-acetylcarbinol with methylamine. This yields the desired isomer L-ephedrine, which is identical in all respects to the alkaloid obtained from ephedra.[8] In view of the difficulties involved in extracting the relatively small concentrations of ephedrine from ephedra, and the fact that the plant serves only as a minor source of the alkaloid anyway, restricting availability of the herb on this basis, although certainly well-intended, seems an excessive measure.[9] The fact that ephedra is commonly abused by consuming excessive amounts for its psychotropic effects is a far better reason for restricting its sale to adults only and limiting the dosage and duration of consumption.

REFERENCES

1. A. Osol and G. E. Farrar Jr. *The Dispensatory of the United States of America,* Second Edition. J. B. Lippincott Company, Philadelphia, Pennsylvania, 1947, pp. 403-407.

2. M. B. Krieg. *Green Medicine.* Rand McNally and Company, Chicago, Illinois, 1964, pp. 415-416.

3. E. Steinegger and R. Hänsel, *Lehrbuch der Pharmakognosie und Phytopharmazie*, Fourth Edition. Springer-Verlag, Berlin, 1988, pp. 451-453.

4. R. Hegnauer. *Chemotaxonomie der Pflanzen,* Volume 1. Birkhäuser Verlag, Basel, Switzerland, 1962, pp. 460-462.

5. M. Weiner. *Health Foods Business* 37(6): 18, 20, 1991.

6. R. E. Stitzel and R. L. Robinson. In *Modern Pharmacology,* Third Edition. C. R. Craig and R. E. Stitzel (Eds.). Little, Brown and Company, Boston, 1990, p. 153.

7. *Lawrence Review of Natural Products,* June, 1989.

8. V. E. Tyler, L. R. Brady, and J. E. Robbers. *Pharmacognosy*, Ninth Edition. Lea and Febiger, Philadelphia, Pennsylvania, 1988, pp. 240-242.

9. Anon. *Health Foods Business* 37(6): 8, 1991.

Evening Primrose

Native to North America, where it is regarded as a noxious weed, the evening primrose *(Oenothera biennis* L.) is considered by some authorities to be a complex of several closely related species. This biennial herb, a member of the family Onagraceae, produces a large number of highly fertile seeds, which are responsible for its introduction and establishment in Europe from ships' ballast in the first years of the seventeenth century. Although the native Indians and early European settlers in America used the whole plant for a variety of conditions, ranging from asthmatic coughs to gastrointestinal disorders to bruises, it is the fatty oil obtained from the small, reddish brown seeds that has caused a resurgence of interest in this herb.[1]

Evening primrose seeds yield about 14 percent fixed oil that, in turn, contains approximately 9 percent of an unusual constituent, *cis*-gamma-linoleic acid (GLA). GLA is a known precursor of prostaglandin E_1, serving as a key intermediate in the biosynthetic pathway leading from *cis*-linoleic acid to that prostaglandin. In fact, conversion of the predominant, essential dietary fatty acid, linoleic acid, to GLA is apparently a limiting step in prostaglandin production. Advocates of the use of evening primrose oil claim that increased intake of it produces a large number of beneficial effects, including, but not limited to, weight loss without dieting, lowered blood cholesterol, lowered blood pressure, cure of rheumatoid arthritis, relief of premenstrual pain, slowed progression of multiple sclerosis, and even the alleviation of hangovers.[2]

Such claims require extensive clinical testing before they can be verified. Scientifically, they would be valid only if all of the specified conditions are favorably influenced by additional production in the body of prostaglandin E_1 and if a deficiency of GLA is the single factor responsible for limited prostaglandin production. Both of these factors remain unproven. If they are not true, then assumptions of the efficacy of evening primrose oil in such conditions is somewhat like assuming one's car will run better if the gas tank is completely full instead of only half full.

Some clinical evidence exists supporting the possible efficacy of evening primrose oil in the treatment of premenstrual syndrome (PMS), mastalgia (sore breasts), multiple sclerosis, atopic eczema, various diabetes-associated problems, cardiovascular disease, rheumatoid arthritis, Sjögren's syndrome, endometriosis, and several other conditions. These studies, which have been reviewed and summarized in some detail, indicate that, at least in Britain, the oil is gaining some medical recognition.[3,4]

However, the validity of some of the reports has also been refuted or at least questioned. An Australian study on the effectiveness of evening primrose oil in treating women with moderate PMS concluded that improvement was solely a placebo effect.[5] Likewise, the methodology of the investigation reporting the utility of the oil in treating atopic eczema has been questioned.[6] At least two clinical trials have shown benefits, particularly in relieving itch for moderate to severa eczema, reducing the need for topical and oral steroids, histamines, and antibiotics. However, two large trials showed no significant benefits.[7] Furthermore, there are no data to support the safety of the long-term consumption of evening primrose oil. However, *cis*-linoleic acid is commonly consumed in the diet, and the amounts of GLA delivered by evening primrose oil corresponds to that produced from normal consumption of *cis*-linoleic acid. Wide availability as a dietary supplement for over fifteen years has resulted in few or no reports of toxicity. The proportions of GLA and *cis*-linoleic acid delivered in human milk is greater than that supplied by evening primrose oil at normal dosage range. When this evidence is taken together, safety seems well established. A potential drug interaction has been identified. Schizophrenic patients receiving phenothiazine epileptogenic drugs should avoid use of evening primrose oil as it could increase risk of temporal lobe epilepsy.[8]

Still, some of the apparently successful results are encouraging and should prompt additional studies.

Evening primrose oil is not inexpensive. A capsule containing 500 mg costs about $0.25, and dosages up to twelve capsules per day have been recommended for certain conditions. Consumers should be aware that some of the products on the market may be adulterated with cheaper oils (soy or safflower), or they may be biologically useless due to decomposition.[9] The finding that seeds of the European black currant, *Ribes nigrum* L., contain about 6 percent GLA, in comparison to less than 2 percent in evening primrose seeds, has now resulted in the appearance on the market of products containing the fixed oil of this plant.[10] Perhaps the richest known plant source of GLA is borage seed, which has been shown to contain up to about 9 percent of GLA.[11] As a result of market demand for less expensive GLA, borage seed oil has recently become available commercially. Unfortunately, the possible presence of toxic pyrrolizidine alkaloids in it may limit its therapeutic utility. (See the monograph on borage for details.)

REFERENCES

1. C. J. Briggs. *Canadian Pharmaceutical Journal* 119: 248-254, 1986.
2. R. A. Passwater. *Evening Primrose Oil.* Kent Publishing, Inc., New Canaan, Connecticut, 1981, 30 pp.
3. A. J. Barber. *Pharmaceutical Journal* 240: 723-725, 1988.
4. C. A. Newall, L. A. Anderson, and J. D. Phillipson. *Herbal Medicines: A Guide for Health-Care Professionals.* The Pharmaceutical Press, London, 1996, pp. 110-113.
5. S. K. Khoo, C. Munro, and D. Battistutta. *Medical Journal of Australia* 153: 189-192, 1990.
6. G. R. Sharpe and P. M. Farr. *Lancet* 335: 1283, 1990.
7. C. A., Newall, L. A. Anderson, and J. D. Phillipson, op. cit., p. 111.
8. Ibid., p. 112.
9. *Lawrence Review of Natural Products* 5(3): 11, 1984.
10. H. Traitler, H. Winter, U. Richli, and Y. Ingenbleek. *Lipids* 19: 923-928, 1984.
11. J. Janick, J. E. Simon, J. Quinn, and N. Beaubaire. Borage: A Source of Gamma Linolenic Acid. In *Herbs, Spices, and Medicinal Plants: Recent Advances in Botany, Horticulture, and Pharmacology,* Volume 4, L. E. Craker and J. E. Simon (Eds.). Oryx Press, Phoenix, Arizona, 1989, pp. 145-168.

Eyebright

Any discussion of this herb must begin with the problems surrounding its name, which are both numerous and difficult. The common name, eyebright, refers to species of *Euphorbia, Lobelia,* and *Sabbatia,* as well as to the plant considered here, *Euphrasia officinalis* L. (family Scrophulariaceae). However, many botanists, particularly those of continental Europe, believe that *E. officinalis* L. represents some four different species and that the plants used medically include *E. rostkoviana* Hayne, *E. stricta* Host, and others. As far as we know, these closely related plants, which do vary slightly in their botanical features, are nevertheless quite similar chemically. It would therefore seem useful, if not entirely accurate, to continue to designate them by the older, more inclusive title, *E. officinalis* L. Currently, 170 species of *Euphrasia* are recognized, forty-six of which occur in Europe.[1]

Eyebright is a small, annual plant with deeply cut leaves, native to the heaths and pastures of Britain, the European continent, and subarctic regions of North America. Similar to several genera of the Scrophulariaceae, Euphrasia is hemiparasitic; the roots have food-gathering nodules that attach to the roots of surrounding plants in order to obtain food. Therefore, it is difficult to cultivate, and virtually the entire supply is harvested from the wild, with little attention to species differences. From July to September, it displays many small, white or purplish flowers variegated with yellow. The various spots and stripes on the flowers cause them to resemble bloodshot, or similarly afflicted, eyes. This, in turn, has caused the plant to be used since the Middle Ages to treat such conditions.[2] The usage was obviously based on the so-called "Doctrine of Signatures."

In his epic poem, *Paradise Lost,* Milton describes how the Archangel Michael used "euphrasy" (eyebright) to clear Adam's sight after his visual nerve had been clouded as a result of eating the "false fruit." This testifies to the popularity of the drug during the latter half of the seventeenth century when Milton was writing his most famous work.

Most modern herbalists recommend a lotion or infusion prepared from the entire overground portion of the plant for conjunctivitis and other eye irritations.[3,4,5] The ancient writers, such as Culpeper and Parkinson, also advised internal consumption of the herb for treatment of similar conditions.[6] Use of the herb continues as a folk medicine, particularly in Eastern Europe, where it is used both topically and internally to treat blepharitis and conjunctivitis, in addition to use as a poultice for styes and eye fatigue.[7]

Chemical studies of eyebright have identified a number of constituents, including aucubin, caffeic and ferulic acids, sterols, choline, various basic compounds, and a volatile oil.[8] However, none of these constituents is known to possess any useful therapeutic properties for the treatment of eye disease, nor are there any modern scientific studies that attempt to measure the effectiveness of the herb. Phenol-carboxylic acids, though, may play a role in perceived antibacterial activity. However, the instillation or application of any nonsterile solution to the eye involves considerable risk of potential infection and should never be advocated or condoned. The practice is particularly hazardous if the nonsterile, homemade lotion contains a large number of principles of unknown safety or efficacy. For this reason, ophthalmic application of eyebright, as advocated by modern herbalists, is definitely not recommended. Little more is known about the drug today than was known in Culpeper's time.

REFERENCES

1. D. J. Mabberly. *The Plant Book*, Second Edition. Cambridge University Press, Cambridge, United Kingdom, 1997, p. 274.

2. M. Grieve. *A Modern Herbal,* Volume 1. Dover Publications, New York, 1971, pp. 290-293.

3. F. and V. Mitton. *Mitton's Practical Herbal.* W. Foulsham and Co., Ltd., London, 1976, p. 86.

4. W. H. Hylton (Ed.). *The Rodale Herb Book.* Rodale Press Book Div., Emmaus, Pennsylvania, 1974, pp. 437-438.

5. M. Stuart (Ed.). *The Encyclopedia of Herbs and Herbalism.* Grosset Dunlap, New York, 1979, p. 189.

6. R. C. Wren and R. W. Wren. *Potter's New Cyclopaedia of Botanical Drugs and Preparations,* New Edition. Health Science Press, Hengiscote, England, 1975, pp. 118-119.

7. *Lawrence Review of Natural Products,* September, 1996.

8. K. J. Harkiss and P. Timmins. *Planta Medica* 23: 342-347, 1973.

Fennel

Because of its pleasant, aromatic odor and its reputation as a stomachic or aid to digestion, fennel is a well-known and widely used folk remedy. The plant, *Foeniculum vulgare* Mill., is a tall (up to five feet), perennial herb with feathery, almost threadlike, leaves and yellow flowers; it is a member of the family Apiaceae. There are numerous cultivated varieties. The dried ripe fruits are the part used for their medicinal virtues, but since they are rather small, they are often referred to as seeds.[1]

Fennel has been recommended for the treatment of a variety of ailments but chiefly as a carminative, that is, an agent which helps expel gas to relieve flatulence. It has often been combined with various purgatives (see senna, for example) to reduce their tendency to cause griping.[2] The drug has a reputation for loosening phlegm and is a common ingredient in cough preparations in Europe. Fennel water is commonly given to infants there to relieve colic and also for its reputed calmative effects.[3]

A large number of constituents have been identified in fennel, but its desirable stomachic and carminative properties are attributed primarily to the volatile oil that exists in the fruits to the extent of about 2 to 6 percent. This oil consists mostly (50 to 90 percent) of *trans*-anethole, with smaller amounts of fenchone (up to 20 percent), estragole, limonene, camphene, and α-pinene. Producing essentially the same actions as the fruit, the oil has been shown to exert spasmolytic (relieves spasms) effects on smooth muscles of experimental animals.[4] This may partially explain fennel's effectiveness as a carminative.

As a folk medicine, fennel has been reported to increase milk secretion, promote menstruation, and increase libido. Estrogenic effects of an acetone extract of the seeds have been observed in male and female rats. Anethole was once thought to be a possible estrogenic component, but more recent research points to anethole polymers including dianethole or photoanethole.[5]

Both fennel fruit and particularly fennel oil are widely employed as fragrance components in a variety of cosmetic preparations and as flavors in foods, beverages, condiments, and the like. There is little question of their safety when used in the very small amounts required for such purposes. Fennel fruit itself, in quantities normally utilized for medicinal teas or similar preparations, is innocuous except for producing a rare allergic response.[6] Fennel volatile oil is quite a different matter. Quantities as small as 1 to 5 ml have caused not only skin irritation but vomiting, seizures, and respiratory problems such as pulmonary edema.[7] For this reason, self-medication with fennel should be restricted to appropriate use of the fruits (seeds); the volatile oil should not be used.

REFERENCES

1. H. W. Youngken. *Textbook of Pharmacognosy,* Sixth Edition. The Blakiston Co., Philadelphia, Pennsylvania, 1948, pp. 614-617.

2. M. Grieve. *A Modern Herbal,* Volume 1. Dover Publications, New York, 1971, pp. 293-297.

3. M. Pahlow. *Das Grosse Buch der Heilpflanzen.* Gräfe un Unzer GmbH, Munich, Germany, 1979, pp. 135-136.

4. A. Y. Leung and S. Foster. *Encyclopedia of Common Ingredients Used in Food, Drugs, and Cosmetics,* Second Edition. John Wiley and Sons, New York, 1996, pp. 240-243.

5. M. Albert-Puleo. *Journal of Ethnopharmacology* 2: 337, 1980.

6. *Lawrence Review of Natural Products,* August, 1994.

7. D. G. Spoerke Jr. *Herbal Medications.* Woodbridge Press, Santa Barbara, California, 1980, p. 70.

Fenugreek

Fenugreek consists of the dried ripe seeds of a small, southern European herb known technically as *Trigonella foenum-graecum* L., a member of the family Fabaceae. It is variously referred to as trigonella or as Greek hayseed. The seeds contain up to 40 percent of a mucilage causing them to be used in various poultices and ointments intended for external application. Fenugreek has also been administered internally for stomach ailments, again due to its soothing mucilaginous properties. Small animal studies have revealed a number of potential therapeutic applications of the seed. These include its use in treating baldness, cancer, elevated cholesterol levels, diabetes, inflammations, microbial and fungal infections, and stomach ulcers.[1] Needless to say, fenugreek's utility for any of these conditions has not been verified in human beings.

In India, fenugreek seeds have traditionally been used as a treatment for diabetes. Various studies have identified hypoglycemic activity of various fenugreek seed extracts in rabbits, rats, and dogs. The effects have been attributed to a number of components, including a defatted seed faction, nicotinic acid, coumarin, and trigonelline. Fenugreek does contain a number of steroidal sapogenins, including yamogenin and diosgenin, which could contribute to some traditional therapeutic applications for the herb. Several small, and mostly uncontrolled, human studies have shown a reduction in plasma glucose concentrations and insulin responses in non-insulin-dependent diabetics. The mechanism of action is not clearly understood.[2] A recent study showed that fenugreek seeds significantly lowered serum cholesterol levels (14 percent reduction) in a twenty-four-week study with sixty non-insulin-dependent diabetics.[3]

The taste of the seed, somewhat reminiscent of maple sugar, accounts for its use as a spice and a flavoring agent, especially in imitation maple syrup.[4] Fenugreek is soothing, flavorful, and even nutritious. Although it is not a particularly potent medicament, it is quite harmless in normal use.

Lydia Pinkham's Vegetable Compound, according to the original formula in her own handwriting, contained fenugreek as its principal ingredient—other than alcohol. Twelve ounces of the seed were combined with 8 ounces of unicorn root (*Aletris farinosa* L.) and six ounces each of life root (*Senecio aureus* L.), black cohosh [*Cimicifuga racemosa* (L.) Nutt.] and pleurisy root (*Asclepias tuberosa* L.) in enough alcohol to make 100 pints of this old-time panacea.[5]

Neither fenugreek nor any of the other ingredients is sufficiently active to account for the remarkable properties attributed to the Vegetable Compound in some of the verses still sung by college students and other irreverent types:

Widow Brown she had no children,
Though she loved them very dear;
So she took some Vegetable Compound,
Now she has them twice a year.

CHORUS

Let us sing of Lydia Pinkham,
And her love for the human race;
How she sells her Vegetable Compound
And the papers publish her face.

REFERENCES

1. *Lawrence Review of Natural Products,* July, 1987.

2. A. Newall, L. A. Anderson, and J. D. Phillipson. *Herbal Medicines: A Guide for Health-Care Professionals.* The Pharmaceutical Press, London, 1996, pp. 117-119.

3. R. D. Sharma, A. Sarkar, D. K. Hazra, B. Misra, J. B. Singh, B. B. Maheshwari, and S. K. Sharma. *Phytotherapy Research* 10: 332-334, 1996.

4. F. Rosengarte, Jr. *The Book of Spices,* Revised Edition. Pyramid Books, New York, 1973, pp. 238-243.

5. S. Stage. *Female Complaints.* W. W. Norton, New York, 1979, p. 89.

Feverfew

Since the time of Dioscorides (78 A.D.), feverfew has been used for the treatment of headache, menstrual irregularities, stomachache, and especially, fevers. In fact, its common name is simply a corruption of the Latin *febrifugia* or fever reducer. The proper scientific name of this strongly aromatic, perennial herb, of the family Asteraceae, is a matter of disagreement among botanists. At different times it has been placed in five different genera! Presently, *Tanacetum parthenium* (L.) Schultz Bip. seems to be the designation most widely accepted in the United States.[1]

In the 1970s, persons unable to obtain relief from the painful symptoms of migraine and arthritis by conventional means began to turn to feverfew as an alternative therapy. Consumption of only two or three fresh leaves daily for prolonged periods was found, for example, to decrease the frequency as well as the pain of migraine attacks. Considerable evidence has now been obtained from studies with fresh whole leaves, freeze-dried powdered leaves, and leaf extracts to confirm feverfew's effectiveness in such cases.[2,3] The principal active constituent of the plant was initially identified as parthenolide, a sesquiterpene lactone.[4] A recent prospective study on the use of feverfew in migraines utilized a dried ethanolic extract prepared from fresh leaf, delivering 0.5 mg of parthenolide on microcrystalline cellulose. The placebo-controlled, crossover study in fifty patients for four months found no significant difference between placebo and treatment groups. Previous clinical studies had used various whole leaf preparations. This calls into question the value of parthenolide as a definitive active constituent; therefore, assurance of parthenolide content may not correspond to clinical efficacy. Additional compounds could work in concert with parthenolide to produce the reduction in frequency and severity of migraine attacks, along with a lessening of migraine-related nausea and vomiting. Clearly more studies are required.[5]

Several sesquiterpene lactones are known to be spasmolytic. That is, they render the smooth muscles in the walls of the cerebral blood vessels less reactive to certain compounds that normally occur in the body and have a pronounced influence on them. Such so-called endogenous substances include norepinephrine, prostaglandins, and serotonin. Thus, the active compound(s) might produce its antimigraine effect in a manner similar to methysergide (Sansert), a known serotonin antagonist.

Regardless of its mode of action, feverfew appears to be a potentially valuable herbal remedy for the treatment of migraine and, possibly, arthritis as well. However, some caution must be observed with respect to the purchase and use of certain commercially available feverfew tablets, some of which have been found to contain only a small percentage of the labeled amount of active plant material.[6] Long-term toxicity tests are urgently needed to establish the herb's safety. It would also be highly desirable to identify all of the active constituents in feverfew and their mechanism of action, as well as to carry out additional clinical tests aimed at discovering the range of modern uses of this ancient herb.[7,8]

REFERENCES

1. C. Hobbs. *HerbalGram* No. 20: 26-36, 1989.
2. A. N. Makheja and J. M. Bailey. *Lancet* II: 1054, 1981.
3. S. Heptinstall, L. Williamson, A. White, and J. R. A. Mitchell *Lancet* I: 1071-1074, 1985.
4. D. V. C. Awang. *Canadian Pharmaceutical Journal* 122: 266-269, 1989.
5. S. Heptinsall and D. V. C. Awang. Feverfew: A Review of Its History, Its Biological and Medicinal Properties, and the Status of Commercial Preparations of the Herb. In *Phytomedicines of Europe Chemistry and Biological Activity*, L. D. Lawson and R. Bauer (Eds.). American Chemical Society, Washington, DC, 1998, pp. 158-175.
6. W. A. Groenewegen and S. Heptinstall. *Lancet* I: 44-45, 1986.
7. V. E. Tyler. *Pharmacy International* 7: 205, 1986.
8. S. Foster. *Feverfew,* Tanacetum parthenium, *Botanical Series No. 310,* Second Edition. American Botanical Council, Austin, Texas, 1996, 8 pp.

Fo-Ti (He-Shou-Wu)

What's in a name? that which we call a rose
By any other name would smell as sweet

William Shakespeare
Romeo and Juliet
Act II, Scene II

Most of the recent herb catalogs list a botanical called fo-ti, sometimes with a cross-reference to he-shou-wu, which many authors insist is the proper title for this drug. Both names refer to the dried tuberous root of *Polygonum multiflorum* Thunb., an herbaceous climbing vine of the family Polygonaceae, native to Japan and widely used as a folk remedy in Chinese medicine.

The reason some writers object to the designation fo-ti for this plant is its potential confusion with the herbal mixture marketed as Fo-ti Tieng.[1] That product, which has a registered trademark name, is totally different from fo-ti or he-shou-wu. Fo-ti Tieng consists of a mixture of the leaves and stems of a diminutive variety of gotu kola [*Centella asiatica* (L.) Urb. of the family Apiaceae], the root of Indian physic [*Gillenia trifoliata* (L.) Moench of the family Rosaceae], which has both emetic and laxative properties, and small amounts of caffeine-containing cola or kolanuts, the dried cotyledons of *Cola nitida* (Vent.) Schott & Endl., family Sterculiaceae, or related species.[2] Since whatever activity the mixture may possess is probably due mostly to the gotu kola contained in it, the reader is referred to the section on that drug in this book for details.

If all of these nomenclatural similarities seem confusing, and they are, keep in mind that these herbal names, and others as well, may have been created on purpose to confuse the consumer into paying a relatively high price for a cheaper but similarly named product. In any event, the one we are discussing here is most commonly called fo-ti, less commonly he-shou-wu, and is the root of *Polygonum multiflorum.*

The Chinese apparently believe that the root exhibits quite different properties according to its size and the age of the plant from which it is derived.[3] Essentially, the older the better: use of 50-year-old root preserves one's natural hair color; 100-year-old root helps one maintain a cheerful appearance; 150-year-old root causes new teeth to grow; 200-year-old root preserves one's youth and energy; and the 300-year-old product makes one immortal. Needless to say, very little (if any) of the truly ancient product is available.

A more realistic appraisal of the use of fo-ti was provided by the American Herbal Pharmacology Delegation in the 1975 report of its visit to the People's Republic of China.[4] That group noted the drug was used alone for scrofula (tuberculosis of the lymph glands), cancer, and constipation and mixed with other medicinals for liver and spleen weakness, vertigo, and insomnia. There is an old report in the European literature mentioning the effectiveness of the drug in treating diabetes,[5] but it is interesting that it is apparently not employed for this purpose in China.

Chemical studies of fo-ti have revealed the presence of chrysophanol and emodin (both in the free state and combined as glycosides) together with a small amount of rhein.[6] All of these anthraquinone derivatives possess cathartic properties that account for the drug's noted effectiveness in the treatment of constipation. In Chinese literature, a distinction is made between "raw fo-ti" and "cured fo-ti." The latter undergoes steaming for twelve hours and sun drying for eight hours, with the processes repeated up to nine times. Recent Chinese works call for steaming for thirty-two hours, which results in a reduction of free and conjugated anthraquiones by 42 to 96 percent. Most Western literature on the drug fails to distinguish between the processed and unprocessed root. More recent Chinese studies have also shown antimicrobial, liver protectant, and cholesterol-reducing activity.[7]

A laxative effect is probably the only real action of fo-ti, at least as far as is presently known. Other species of *Polygonum* do contain leucoanthocyanidins that possess anti-inflammatory activity, decrease blood coagulability, and have various cardiovascular effects.[8] It is possible that some of these compounds may eventually be discovered in fo-ti. Until then, the drug must be categorized simply as a laxative with various undetermined side effects.

REFERENCES

1. D. Mowrey. *Herbalist* 5(1): 14-15, 1980.

2. V. E. Tyler, L. R. Brady, and J. E. Robbers. *Pharmacognosy,* Ninth Edition. Lea and Febiger, Philadelphia, Pennsylvania, 1988, pp. 427-473.

3. C. Lam. *Herbalist* 2(7): 12-14, 1977.

4. *Herbal Pharmacology in the People's Republic of China.* National Academy of Sciences, Washington, DC, 1975, p. 186.

5. O. A. F. Gnadt. *Die Pharmazie* 1: 103-107, 1946.

6. P. H. List and L. Hörhammer (Eds.). *Hagers Handbuch der Pharmazeutischen Praxis,* Fourth Edition, Volume 6A. Springer-Verlag, Berlin, 1977, p. 823.

7. A. Y. Leung and S. Foster. *Encyclopedia of Common Natural Ingredients Used in Food, Drugs, and Cosmetics,* Second Edition. John Wiley and Sons, New York, 1996, pp. 250-253.

8. *Herbal Pharmacology in the People's Republic of China,* op. cit., 1975, p. 186.

Garcinia

Garcinia is the genus name of a group in the family Guttiferae with upwards of 200 species of slow-growing trees and shrubs native to Old World tropics, especially Asia, Polynesia, and South Africa. Garcinia is also a common herb trade name that, in the American market, has come to refer to products derived from the fruits of *Garcinia cambogia* Desr. About thirty *Garcinia* species are found in India, including *G. cambogia*, which is a small or medium-sized tree with a rounded crown and shining dark green leaves up to five inches long and three inches wide. The oval, fleshy fruits, about two inches in diameter, turn yellowish or red when ripe. In India, the tree grows in evergreen forests up to an altitude of 6,000 feet. The tree has been called "Malabar tamarind," a misnomer, as the fruit should not be confused with tamarind (*Tamarindus indica* L.). The fruits of the two trees have a similar acidic flavor, typical of citric acid.[1,2]

The fruits are common in markets of Kerala State in South India and have long been used as a condiment in India, especially the West Coast, as well as by Muslim populations in Laos, Malaysia, Thailand, and Myanmar (formerly Burma). In Muslim traditions in India, the dried rind has been used as a flavoring condiment, especially in curries, as a substitute for limes or tamarind. Although the fruits of *G. cambogia* are edible, they are generally considered to be too acidic to be eaten raw. The dried fruit is concentrated for use in curries, with an acid ratio up to 60 percent. An extract of the dried fruits is traditionally taken after lamb dishes with a high fat content to aid in digestion. In Sri Lanka, the fruits are harvested just before ripening, then the thick-fleshed fruit is cut into sections, sun dried, and stored for future use. Traditionally, it is used along with salt for curing dried fish. Commercially this process is known as "Colombo curing." In folk traditions in India, the dried fruit rind has been decocted for rheumatism and bowel complaints. [3]

In 1965, Indian researchers identified a new compound (−)-hydroxycitrate [(−)-HCA] as the principal acid in the fruits of

G. cambogia. Previously, the organic acids present in the fruit had been erroneously identified as tartaric and citric acids.[4] (−)-HCA is found in the fruits of two other *Garcinia* species, *G. indica* (Thouras) Choisy and *G. atriviridis* Griff. The highest level is found in *G. cambogia* fruits at up to 16 percent in the dried fruit.

Studies in the early 1970s showed that (−)-HCA inhibits citrate cleavage enzyme and fatty acid synthesis in vivo in the rat liver.[5] Further work revealed that (−)-HCA decreased the rate of lipogenesis (when administered intravenously or interperitonally), and oral administration reduced fatty acid and cholesterol synthesis only when given before feeding. This lead to the supposition that if (−)-HCA could inhibit lipogenesis, but not affect normal energy production by cell mitochondria in humans, it could be of value in treating certain lipid disorders and obesity.[6]

Up until the early 1990s, numerous studies were published on the mechanism of action of (−)-HCA on fatty production and metabolism in rats. In various studies, (−)-HCA was found to produce significant reduction in food intake and body weight gain in rats. It was found to be a potent competitive inhibitor of ATP citrate lyase, an enzyme involved in the conversion of carbohydrates to fat in mammals. A number of studies have shown that the compound significantly reduced fatty acid synthesis in the liver and adipose tissue of laboratory rats, while decreasing weight gain in various lean and obese rat experimental models.[7,8] The compound also produced an, as yet, not completely understood appetite suppressant effect. One hypothesis that has been put forward is whether the anorexia could be produced by alteration of metabolic processes resulting from diversion of dietary carbohydrates and their metabolites to lipid synthesis. One study also found that in obese rats, although (−)-HCA produced a reduction in food intake and body weight, the percentage of body fat and the size of fat cells were not reduced by the compound during the feeding period. In other words, the effect of (−)-HCA was different in lean rats and obese rats.[9]

Another problem with garcinia is that the (−)-HCA in it occurs almost entirely in the lactone form. This must be converted to the acid form before it can become active. There is, at present, no evidence that this occurs with any degree of efficiency in the human body following consumption.[10]

Many products in the weight-loss realm are directed toward taking in fewer calories or stimulating faster burning of calories. Unlike plant-based central nervous system stimulant products seen in weight-loss formulations, such as ephedra (ma huang) or caffeine-containing products, *Garcinia* does not produce stimulation of the central nervous system, hence enhanced calorie-burning. *Garcinia cambogia* is an ingredient that utilizes the approach of altering lipid and carbohydrate metabolism. Acute and chronic toxicity studies on garcinia or its constituents are generally lacking from the scientific literature. Despite a relatively large number of pharmacological studies, most involving rats, clinical studies on the effects of garcinia preparations on obese humans are noticeably absent from the literature. A recent book does cite data on safety and human clinical experience, however, without citations to specific studies.[11] New studies are certainly needed before garcinia or its constituents can be considered useful for controlling obesity in humans. In the meantime, garcinia products have successfully lightened the bank accounts of consumers.

REFERENCES

1. Anon. *Wealth of India,* Volume IV, Council of Scientific and Industrial Research, New Delhi, 1956, pp. 99-100.

2. S. Foster. *Health Foods Business,* June, 27, 1994.

3. Anon. Op. cit., pp. 99-100.

4. Y. S. Lewis and S. Neelakantan. *Phytochemistry* 4: 619-625, 1965.

5. J. H. Lowenstein. *The Journal of Biological Chemistry* 346(3): 629-632, 1971.

6. A. C. Sullivan, J. G. Hamilton, O. N. Miller, and V. R. Wheatley. *Archives of Biochemistry and Biophysics* 150: 183-190, 1972.

7. A. C. Sullivan and J. Triscari. *The American Journal of Clinical Nutrition* 30(5): 767-776, 1977.

8. A. C. Sullivan, J. Triscari, and H. E. Spiegel. *The American Journal of Clinical Nutrition* 30(5): 777-784, 1977.

9. S. Foster, op. cit.

10. D. Clouatre and M. Rosenbaum. *The Diet and Health Benefits of HCA (Hydroxycitric Acid).* Keats Publishing, New Canaan, Connecticut, 1994, pp. 8-11.

11. M. Maheed, R. Rosen, M. McCarty, A. Conte, D. Patil, and E. Butrym. *Citrin®—A Revolutionary, Herbal Approach to Weight Management.* New Editions Publishing, Burlingame, California, 1994.

Garlic and Other Alliums

Even relatively well-informed readers, asked to name the most popular herbal panacea or cure-all, might be inclined to say ginseng. They would be wrong. As broad as its claims of curative properties are, ginseng's hypothesized range of therapeutic and overall use, until recently limited primarily to Asia, does not begin to compare with the many and varied worldwide applications of garlic and its near relatives, onions, leeks, and shallots. These well-known members of the lily family (Liliaceae) all belong to the genus *Allium:* garlic is *A. sativum* L.; onion is *A. cepa* L.; the leek is *A. ampeloprasum* L.; and the shallot is *A. ascalonicum* L.

The bulbs, and occasionally the leaves, of these plants, designated collectively as alliums, have been used by people since the earliest days of recorded history as both food and medicine. Keller has listed some 125 different uses of the alliums, only one of which is for culinary purposes.[1] Some of the others are very broad categories indeed, such as, cures "all diseases. Others are contradictory: cures "hypertension and "low blood pressure. Still others are more mythical than medical; garlic, for example, was noted for its ability to ward off vampires, demons, witches, and similar imaginary beings. Above all, the alliums were valued as aphrodisiacs, agents that produce sexual desire and improved performance.

Most of the modern folkloric medicinal use of these plants has focused on garlic. Although some advocates still recommend its use for everything from cancer and tuberculosis to hemorrhoids and athlete's foot,[2,3,4] the most frequent use of garlic in recent times has been in treating atherosclerosis and high blood pressure. A prominent secondary application is to provide relief from various stomach and intestinal ailments.[5]

The chemistry of garlic has been extensively investigated. Its bulbs contain an odorless, sulfur-containing amino acid derivative known as alliin (*S*-allylcysteine sulfoxide). This parent substance has no antibacterial properties, but when the bulbs are ground, alliin comes into contact with the enzyme alliinase, which converts alliin to allicin (allyl 2-propenethiosulfinate), a potent antibacterial agent. Unfortunately, allicin is extremely odoriferous; it is the carrier of the typical garlic odor. It is also unstable; when garlic bulbs are subjected to steam distillation to obtain their volatile oil (0.1 to 0.36 percent), some of it breaks down to yield diallyldisulfide and related garlic-smelling compounds.[6] Still, the allicin with its antibacterial activity against numerous gram-positive and gram-negative pathogenic organisms may account for the alleged effectiveness of garlic in the folk treatment of infectious conditions.

Beneficial effects of garlic on digestive disturbances and on the cardiovascular system have been demonstrated in animals and in human beings. Experiments showed that garlic inhibited experimental hypercholesterolemia (high blood cholesterol) in rabbits and reduced high blood pressure in both dogs and people. Small doses increased the tonus of the smooth muscles of the intestines, thereby increasing peristalsis, but large doses inhibited such movements.[7]

Results of some laboratory studies as well as two extensive tests in human beings have given some indications that garlic may prevent the development of cancer by stimulating the immune system and hindering the growth of malignant cells. Two large-scale studies, one in China and one in Italy, found fewer stomach cancer deaths among people who regularly consumed large amounts of alliums.[8] These preliminary indications are certainly worthy of further research.

The ability of garlic to provide some protection against atherosclerosis, coronary thrombosis, and stroke is believed to be related, at least in part, to its ability to inhibit aggregation of the blood platelets. This property, initially attributed to allicin, is now known to be due to a compound designated ajoene, a self-condensation product of allicin.[9,10] As an antithrombotic (clot-preventing) agent, ajoene is at least as potent as aspirin, and its activity is enhanced by two breakdown products which accompany it in garlic and which are also mildly antithrombotic.

In 1989, Dutch investigators published a detailed analysis of all the clinical studies that had been carried out on the effects of garlic and onions on various cardiovascular risk factors in human beings.

They concluded that with respect to fresh garlic, claims were valid for beneficial effects on blood cholesterol levels, fibrinolytic activity, and platelet aggregation—but only at relatively high dosage levels. Most studies involved ingestion of 0.25 to 1 g of fresh garlic per kilogram of body weight per day. This would be equivalent to a range of from five to twenty average-sized (4 g) cloves of fresh garlic daily for a 175-pound person. Results obtained from studies with commercial garlic preparations and with fresh onions were contradictory and did not allow conclusions to be drawn with respect to their efficacy.

In addition, they found that all trials published prior to 1989 showed severe methodological shortcomings. Some were not randomized nor blinded, and the numbers involved in others were small. It was concluded that inadequate scientific justification exists for garlic supplementation of the diet.[11] However, it must be noted that under the strict guidelines applied by these critics, it would not be possible to justify the consumption of any drug that lowered cholesterol or inhibited platelet aggregation because of their belief that "it is doubtful whether healthy people need a change of their fibrinolytic activity or inhibition of their platelet aggregation. The truly important conclusion that may be reached from this highly critical analysis is that fresh garlic, in appropriate amounts, does produce both of these effects, in addition to lowering cholesterol.

In the 1990s, there has been an explosion in garlic research. A recent review by an American garlic specialist revealed that from 1990 to 1996, a total of 1,158 pharmacological studies had been published, 208 of which were human studies. Studies have focused on four major areas of research, including cardiovascular effects (334 studies), anticancer (221 studies), antimicrobial (252 studies), and antioxidant effects (60 studies). In 1994, Americans consumed 2.5 g of garlic per day, for a total consumption of 438 million pounds.[12]

The most comprehensive review of the scientific literature ever published on garlic, including the numerous clinical studies of the past decade, contains a revised analysis of dosing of garlic and its preparations. Based on the latest clinical studies, a significant lipid-lowering effect can be expected with a daily dose of garlic powder corresponding to 600 to 900 mg of garlic powder containing 1.3

percent alliin or 0.6 percent allicin, delivering a daily dose of 3.6 to 5.4 mg allicin.[13]

A brief look at the stability of the active principles in garlic will help us understand why equivocal results were obtained with the commercial preparations. These include garlic that has been dried by various means (heated or freeze-dried), distilled or extracted garlic oils, aged garlic, as well as garlic that has been deodorized by unspecified processes. Dried garlic preparations contain neither allicin, the antibacterial principle, nor ajoene, the anticlotting agent. They do contain alliin, the inodorous precursor of both. However, to convert alliin to allicin and, ultimately, ajoene requires the action of the enzyme alliinase. It may be present in garlic that has been dried at a relatively low temperature, but it is unstable in the presence of acids. So, when dried garlic is consumed and reaches the stomach, the enzyme tends to be destroyed by the acid there, and little conversion to the active principles takes place.[14]

Therefore, dried garlic preparations are most effective if they are enteric coated, that is, protected by a coating that allows them to pass intact through the stomach to the small intestine where the enzyme can act under the alkaline conditions present there. Such enteric-coated capsules or tablets of freeze-dried garlic should theoretically be more effective than even fresh garlic in which alliinase is also destroyed in the stomach. Fresh garlic, therefore, releases its active principles mainly in the mouth during chewing, not later in the stomach. This does raise the intriguing possibility that the effectiveness of fresh garlic could be greatly increased if it were thoroughly chewed and held in the mouth for a longer period prior to swallowing. However, to most, this is probably a rather distasteful procedure.

At first glance, it would seem to be possible to judge the effectiveness of any garlic preparation by taking it, waiting to see if a typical garlic taste develops, and then estimating the intensity of the taste. However, this is not necessarily the case with enteric-coated preparations. Allicin released in the small intestine will react rapidly and spontaneously with the amino acid cysteine derived from proteinaceous food consumed with the garlic. The *S*-allylmercaptocysteine thus formed effectively binds the odoriferous allicin and prevents it from reaching the bloodstream, as such. It is thus not possible to rely on taste as a measure of the effectiveness of such garlic preparations.

Consumption of moderate quantities of garlic, even on a daily basis, should not pose any particular health risk for normal persons. Larger amounts (five or more cloves per day) can result in heartburn, flatulence, and related gastrointestinal problems. Some persons may suffer allergic reactions as well. Consumption of garlic reduces the clotting time of the blood, which may cause medical problems in certain individuals.[15] Persons taking aspirin or other anticoagulant drugs should avoid eating large amounts of the herb.

At present, the best way to ensure the effectiveness of garlic for any of the conditions for which it may be helpful is to eat it raw in relatively large quantities.[16] However, following this advice may reduce the number of one's close associates in direct proportion to the quantity and frequency of the alliums consumed. Consumption of enteric-coated tablets of a suitably prepared garlic powder appears to be a viable, relatively odor-free alternative.

Based on current knowledge, there does seem to be more truth than poetry in the old Welsh rhyme:

> Eat leeks in March and wild garlic in May,
> And all the year after physicians may play.

REFERENCES

1. M. S. Keller. *Mysterious Herbs & Roots.* Peace Press, Culver City, California, 1978, pp. 162-211.
2. T. Watanabe. *Garlic Therapy.* Japan Publications, Tokyo, 1974.
3. G. J. Binding. *About Garlic.* Thorsons Publishers Ltd., Wellingborough, England, 1970.
4. P. Airola. *The Miracle of Garlic.* Health Plus, Phoenix, Arizona, 1978.
5. S. Foster. *Allium sativum, Garlic Botanical Series No. 311,* Second Edition. American Botanical Council, Austin, Texas, 1996, 7 pp.
6. E. Steinegger and R. Hänsel. *Lehrbuch der Pharmakognosie und Phytopharmazie,* Fourth Edition. Springer-Verlag, Berlin, 1988, pp. 633-637.
7. S. Foster, op. cit.
8. Anon. *Consumer Reports Health Letter* 3: 17, 19-20, 1991.
9. E. Block, S. Ahmad, M. K. Jain, R. W. Crecely, R. W. Apitz-Castro, and M. R. Cruz. *Journal of the American Chemical Society* 106: 8295-8296, 1984.
10. E. Block. *Scientific American* 252(3): 114-119, 1985.

11. J. Kleijnen, P. Knipschild, and G. Ter Riet. *British Journal of Clinical Pharmacology* 28: 535-544, 1989.

12. L. Lawson. Garlic: A Review of Its Medicinal Effects and Indicated Active Compounds. In *Phytomedicines of Europe Chemistry and Biological Activity,* L. D. Lawson and R. Bauer (Eds.). American Chemical Society, Washington, DC, 1998, pp. 176-209.

13. H. P. Koch and L. D. Lawson. *Garlic: The Sciences and Therapeutic Application of Allium sativum L. and Related Species,* Second Edition. Williams and Wilkins, Baltimore, Maryland, 1996.

14. C. Friedl. *Zeitschrift für Phytotherapie* 11: 203, 1990.

15. K. D. Rose, P. D. Croissant, C. F. Parliment, and M. B. Levin. *Neurosurgery* 26: 880-882, 1990.

16. H. P. Koch. *Deutsche Apotheker Zeitung* 129: 1991-1997, 1989.

Gentian

Bitter substances taken before eating are supposed to improve both the appetite and the digestion by increasing the flow of gastric juices. Although there is little evidence to support this view in normal, healthy persons, it is probably factual for those suffering from conditions such as anemia.[1] Nevertheless, many well people genuinely believe in the beneficial effects of so-called bitter stomachics.

The use of such old-time proprietary remedies as Hostetter's Celebrated Stomach Bitters in the United States has largely given way to adding a few drops of Angostura Bitters (which contains gentian, not angostura) to the evening cocktail. In Europe, various alcoholic beverages flavored with gentian and other bitter principles remain extremely popular and are widely consumed, especially before eating a heavy, fatty meal that may cause digestive difficulties. My friends in Germany recommend it and use it themselves before attacking a large, spit-roasted *Schweinshaxen* (pork shank) at a specialty restaurant such as the Haxnbauer in Munich.

Gentian, the dried rhizome and roots (underground parts) of *Gentiana lutea* L. of the family Gentianaceae, a moderately tall, perennial herb with an erect stem and large, ovate leaves, is by far the most popular of the bitter stomachics. The plant has large flowers that grow in characteristic orange-yellow clusters. It is a native of the alpine and subalpine pastures of central and southern Europe and is extensively cultivated there.[2] The bitter alcoholic beverage prepared from it has become almost a trademark of specific regions in several European countries.

Modern herbalists extol the virtues of gentian far beyond those of a simple bitter. It is believed to be useful in the treatment of exhaustion from chronic disease and in cases of general debility as well. They view it as a strengthener of the human system—in other words, as a tonic. It is also said to be useful as a febrifuge (reduces fever), emmenagogue (stimulates the menstrual flow), anthelmintic (expels intestinal worms), and antiseptic. In their view, it is helpful in treating hysteria and, in combination with other drugs, malaria.[3] Gentian is usually consumed in the form of a tea or as one of the commercially available alcoholic extracts.

Glycosides known as amarogentin and gentiopicrin are primarily responsible for the bitter taste of gentian. In addition, the plant contains several alkaloids (mainly gentianine and gentialutine), xanthones, triterpenes, and sugars. Aside from its action as a bitter stomachic, none of the other purported effects is well documented in human beings. Some experiments on small animals indicate that gentian may increase the secretion of bile; the alkaloid gentianine also exhibits anti-inflammatory properties.[4]

Widespread use of gentian as an appetite stimulant and digestive aid would seem to favor the herb as effective for these conditions. However, since it is normally consumed as an alcoholic beverage, it is difficult to separate the effects of gentian from those of alcohol, which are very similar, at least when the alcohol is consumed in moderate amounts.[5] In normal individuals, gentian is unlikely to produce undesirable side effects; however, occasional headaches have been reported. Overdoses (greater than 0.1 to 2 g of the drug decocted in 150 ml of water, three times per day), could lead to nausea or vomiting. Use is contraindicated in gastric or duodenal ulcers and hyperacidity.[6] Pahlow warns that the drug may not be tolerated well by those with very high blood pressure or by expectant mothers.[7] Actually, these people should be very cautious about using any medication, herbal or otherwise.

REFERENCES

1. T. Sollmann. *A Manual of Pharmacology,* Seventh Edition. W. B. Saunders, Philadelphia, Pennsylvania, 1948, pp. 167-168.

2. H. W. Youngken. *Textbook of Pharmacognosy,* Sixth Edition. The Blakiston Co., Philadelphia, Pennsylvania, 1948, pp. 670-674.

3. M. Grieve. *A Modern Herbal,* Volume 1. Dover Publications, New York, 1971, pp. 347-349.

4. A. Y. Leung and S. Foster. *Encyclopedia of Common Natural Ingredients Used in Food, Drugs, and Cosmetics,* Second Edition. John Wiley and Sons, New York, 1996, pp. 267-269.

5. T. Sollmann, op. cit., pp. 615-616.

6. ESCOP. Gentianae Radix. In *Monographs on the Medicinal Uses of Plant Drugs,* Volume 4. European Scientific Cooperative on Phytotherapy, Exeter, United Kingdom, 1997.

7. M. Pahlow. *Das Grosse Buch der Heilpflanzen.* Gräe and Unzer GmbH, Munich, Germany, 1979, pp. 124-126.

Ginger

Ginger is technically a rhizome (underground stem) of the plant *Zingiber officinale* Roscoe of the family Zingiberaceae. In commerce it is frequently referred to as Jamaica ginger, African ginger, or Cochin ginger, according to its geographical origin. Ginger was known in China nearly 2,500 years ago, and it continues to be valued throughout the world as a spice or flavoring agent.[1] However, it also has the reputation, particularly in Asian medicine, as a carminative (digestive aid), stimulant, diuretic, and antiemetic.

The characteristic aroma of ginger is due to a volatile oil, which it contains in amounts of about 1 to 3 percent. Its pungency is attributed to ginger oleoresin (mixture of volatile oil and resin). Components of this oleoresin known as gingerols have recently been studied and found to possess cardiotonic as well as antipyretic, analgesic, antitussive (anticough), and sedative properties when administered to laboratory animals.[2]

Over fifteen years ago, a double-blind study conducted on thirty-six college students with a high susceptibility to motion sickness concluded that 940 mg of powdered ginger was superior to 100 mg of dimenhydrinate in reducing symptoms when consumed twenty-five minutes prior to tests in a tilted rotating chair.[3] Five additional studies have been made since that time by other investigators; two yielded negative results, but three confirmed the initial findings.[4] German health authorities have subsequently concluded that ginger, at an average daily dose level of 2 to 4 g, is effective for preventing motion sickness and is also useful as a digestive aid.[5] Ginger is also deemed useful as a postoperative antiemetic for minor outpatient surgical procedures.[6] Any antiemetic effects of ginger are due to its local action in the stomach, not to any central nervous system activity.

Ginger is ordinarily taken in the form of capsules, each containing 500 mg of the powdered herb. It may also be consumed as a tea or in the form of candied ginger, which is readily available in Asian food markets. There are no reports of severe toxicity in humans from eating ginger, but some of the recent pharmacological studies of its constituents would seem to indicate that very large overdoses might carry the potential for causing depression of the central nervous system and cardiac arrhythmias. In the meantime, further investigations of the chemical constituents and the therapeutic properties of ginger are certainly warranted.

REFERENCES

1. V. E. Tyler, L. R. Brady, and J. E. Robbers. *Pharmacognosy,* Ninth Edition. Lea and Febiger, Philadelphia, Pennsylvania, 1988, p. 150.

2. *Lawrence Review of National Products,* April, 1986.

3. D. B. Mowrey and D. E. Clayson. *Lancet* I: 655-657, 1982.

4. S. Holtmann, A. H. Clarke, H. Scherer, and M. Höhn. *Acta Otolaryngologica* 108: 168-174, 1989.

5. *Bundesanzeiger,* May 5, 1988, March 6, 1990.

6. ESCOP: Zigiberis Rhizoma. In *Monographs on the Medicinal Uses of Plant Drugs*, Volume 1. European Scientific Cooperative on Phytotherapy, Exeter, United Kingdom, 1996.

Ginkgo

The ginkgo tree, *Ginkgo biloba* L., is the last remaining member of the Ginkgoaceae, a family that once numbered many species. Having survived unchanged in China for more than 200 million years, it was brought to Europe in 1730 and since then has become a popular ornamental tree in parks and gardens throughout the world. Because of its hardiness, it even thrives along the heavily trafficked streets of some of our major cities.[1]

Ginkgo's fleshy seeds have been valued in China for their medicinal properties since 2800 B.C., but it is only in the last forty years that the leaves of this living fossil have been utilized extensively in Western medicine. Unlike many of the herbs in use today, ginkgo leaves are not used so much in their crude state as in the form of a concentrated, standardized ginkgo biloba extract (GBE). This extract has become a very popular drug in Europe, where it is widely used for its beneficial effects on the circulatory system.[2] In fact, during 1988, physicians in Germany wrote more prescriptions (5.4 million) for GBE than for any other drug. It is also available there as an over-the-counter (OTC) drug.[3]

GBE is produced from green-picked leaves grown on plantations in the United States, France, Japan, and South Korea that have been specifically developed for pharmaceutical purposes. After drying and milling, the leaves are extracted with an acetone-water mixture under partial vacuum. The organic solvent is then removed and the extract processed, dried, and standardized. GBE is then adjusted to a potency of 24 percent flavonoids (mostly flavonoid glycosides and quercetin) and 6 percent terpenes (principally a unique group of diterpenes known as ginkgolides, composed of 2.9 percent bilobalide and 3.1 percent ginkgolides A, B, C, and J). In the extraction process, other constituents are removed including ginkgolic acid, an allergen.[4,5] The product is marketed in both solid and liquid form; each tablet or capsule usually contains 40 mg of the extract.

There is an impressive body of literature attesting to the effectiveness of GBE in treating ailments associated with decreased cerebral blood flow, particularly in geriatric patients.[6] These conditions include short-term memory loss, headache, tinnitus, depression, and the like. Clinical and pharmacological studies have shown that GBE promotes vasodilation and improved blood flow both in the arteries and capillaries. There are also indications that it is an effective free-radical scavenger. Large doses are required, which explains why a concentrate is used rather than the herb itself. GBE does reduce the clotting time of blood, which may be of concern to those already taking anticoagulants. Very large doses may cause restlessness, diarrhea, nausea, vomiting, and other unpleasant effects, usually of a relatively mild nature. If these occur, cease taking the drug or reduce the dosage.

In 1989, the methodology of forty-eight of the studies supporting the effectiveness of ginkgo was criticized by a group of German clinical pharmacologists.[7] This criticism and responses by those defending the validity of the studies created something of a cause célèbre in the European medical literature.[8] A careful reading of the articles reveals that the criticism was not well substantiated.[9] Although some of the studies may have been deficient in one or more particulars, this is true of most clinical investigations. It must be concluded that the critics' standards were somewhat unrealistic and the reports of the effectiveness of GBE are, in general, valid.

To date, more than forty clinical trials have been published on the use of GBE for cerebral insufficiency, ten to twelve of which have been deemed methodologically sound. In Germany, GBE is licensed for treating cerebral dysfunction symptoms (such as memory difficulties, dizziness, tinnitus, headaches, and emotional stability with anxiety). It is also used in the supportive treatment of intermittent claudication and hearing loss due to cervical syndrome.[10]

Until now, most clinical studies have been conducted in Europe. The first large-scale American clinical study on GBE, conducted at six research centers, was published in late 1997. The trial, which lasted for fifty-two weeks, started with 327 patients, but in the end, data were available for 202 patients. Patients admitted to the study were diagnosed with either Alzheimer-type dementia or multi-infarct dementia. Patients received 120 mg per day (40 mg t.i.d.). The authors concluded that the *Ginkgo biloba* leaf extract, EGb 761, was both safe and stabilizing, and in a significant number of patients, it improved cognitive performance and social functioning. More studies are needed to determine the effects of long-term administrations of the extract and what the effects might be of removal of the treatment. Several pilot studies have also suggested that higher dosages may be more effective. Further research will have to be conducted to resolve these and other questions.[11]

Because the present regulations in this country pertaining to herbal products are less than realistic, GBE is presently marketed here only as a dietary supplement. It is worth noting that almost all of the scientific and clinical studies on the effectiveness of GBE have been carried out on two standardized extracts, cited in the literature as EGb-761, originally prepared by the Willmar Schwabe company in Germany, or another designated LI 1370, produced by Lichtwer Pharms, also in Germany. Both of these extracts are marketed in the United States by those organizations or their subsidiaries.[12] If consumers want well-tested products, they will need to check the origin of the GBE product that they are contemplating purchasing.

REFERENCES

1. W. Caesar. *Deutsche Apotheker Zeitung* 129: 2430-2431, 1989.
2. H. Schilcher. *Zeitschrift für Phytotherapie* 9: 119-127, 1988.
3. Anon. *Lancet* II: 1513-1514, 1989.
4. K. Drieu. Preparation and Definition of Ginkgo Biloba Extract. In *Rökan: Ginkgo Biloba,* E. W. Fünfgeld (Ed.). Springer-Verlag, Berlin, 1988, pp. 32-36.
5. B. Ahlemeyer and J. Krieglstein. Neuroprotective Effects of Ginkgo Biloba Extract. In *Phytomedicines of Europe Chemistry and Biological Activity,* L. D. Lawson and R. Bauer (Eds.). American Chemical Society, Washington, DC, 1988, pp. 210-220.
6. H. Schilcher. *Zeitschrift für Phytotherapie* 9: 119-127, 1988.
7. P. S. Schönhöfer, H. Schulte-Sasse, C. Manhold, and B. Werner. *Internistische Praxis* 29: 585-601, 1989.
8. U. Stein. *Lancet* 335: 475-476, 1990.
9. W.-U. Weitbrecht. *Internistische Praxis* 30: 621-622, 1990.
10. B. Ahlemeyer and J. Krieglstein. In op. cit., p. 217.
11. L. P. Le Bars, M. M. Katz, N. Berman, T. M. Itil, A. M. Freedman, and A. F. Schatzberg. *Journal of the American Medical Association,* 278: 1327-1332, 1997.
12. S. Foster. *Ginkgo,* Ginkgo biloba, *Botanical Series No. 304*, Second Edition. American Botanical Council, Austin, Texas, 1996, 8 pp.

Ginseng and Related Herbs

That too much has been written about ginseng is a gross understatement. The Research Institute of the Office of [Ginseng] Monopoly, Republic of Korea, has cited and abstracted 1,191 books and papers published on the herb between 1687 and 1975.[1] And since 1975, the volume of writing has increased enormously. To say that we now have a reasonable knowledge of the botany and chemistry of ginseng is a fact. Recent studies in both areas have made significant contributions to our knowledge. But to say that we have an adequate understanding of the way ginseng works (if indeed it does) in helping to maintain health or in preventing or curing disease in the human body is an enormous exaggeration. Most of the literature in this area is based more on superstition and subjective opinion than on objective, scientific evidence.

Nothing about ginseng seems to be totally free from controversy. Even the proper name of the low-growing, shade-loving perennial herbs of the family Araliaceae that yield the highly valued roots is not agreed upon by all specialists. Following American authorities,[2] we are using the designation *Panax ginseng* C. A. Mey. to designate the species widely cultivated and utilized in Asia; it is also known as *P. schinseng* Nees.

The native American species that occurs more or less commonly in wooded areas from Quebec to Minnesota and south to Georgia and Oklahoma is known as *P. quinquefolius* L. It is so intensively sought in the United States that it has been declared an endangered species. Its collection and sale are subject to registration, permits, reports, a collector-education program, and an official ginseng season. Unfortunately, the legal protections are seldom enforced.[3] The plant is also cultivated in the United States, but it is a slow-growing and exacting crop requiring about six years to produce a marketable root. Commercial cultivation is largely limited to Wisconsin, Ontario, and British Columbia. Recently, American ginseng has begun to be grown extensively in China. In the Chinese system of medicine, its properties are said to be "cold," whereas those of regular Asian ginseng are "hot." This causes the American product to be viewed as superior for the treatment of certain conditions.

Basically, ginseng has been used for centuries in Asia, especially in China, Japan, Korea, and parts of the former Soviet Union, as a cure-all or panacea. This usage no doubt stems from the "Doctrine of Signatures" because the root is often decidedly manlike in appearance and therefore useful in the treatment of all "man's afflictions." Farnsworth has pointed out that the widest use of the drug in those areas is not in curing a particular disease but in a supportive role to maintain health.[4] In this regard, ginseng is analogous to the ubiquitous vitamin tablets here, but with one important addition: the drug is commonly believed to have a favorable influence on sexual potency, in other words, to be an aphrodisiac.

More recently, as ginseng consumption has galloped across the Western world, its adaptogenic effects are being emphasized. As an adaptogen, it is believed to produce a state of increased resistance of the body to stress, overcoming disease by building up our general vitality and strengthening our normal body functions.[5] Although such indirect effects are naturally somewhat difficult to verify scientifically, favorable modification by ginseng of the stress effects of temperature changes, diet, restraint, exercise, and the like have been recorded. Moreover, useful pharmacologic effects in such conditions as anemia, atherosclerosis, depression, diabetes, edema, hypertension, and ulcers have also been documented.[6] In one area, however, ginseng's purported beneficial effects remain unsubstantiated: there is no scientific evidence of enhanced sexual experience or potency resulting from its use.[7]

The principles believed to be responsible for ginseng's activities are triterpenoid saponins that exist in the root in large numbers.[8] Again, the nomenclature of these compounds is extremely confusing and complex, for some of the same ones were isolated by different groups of investigators and given different names. Then, too, there are differences in composition between the Asian and American ginseng species. The active saponins are called ginsenosides by Japanese and panaxosides by Russian scientists. Thus, we have at least eighteen saponins in Asian ginseng, such as ginsenoside R_c, which is the same as panaxoside D. The same compound isolated from American ginseng is also known as panaquilin C.[9] Confusing? Of course! So leave the details to the scientists and simply remember that whatever pharmacological activity ginseng may possess is probably due mainly to its many chemical compounds, which are triterpenoid saponins.

Obtaining the authentic ginseng product is a problem. Quality root is extremely expensive—the best grades of Korean Red (a specially "cured" root) retail at more than $20 an ounce. This relatively high cost plus lack of quality control in many areas of the health food industry have resulted in commercial ginseng products (teas, powders, capsules, tablets, extracts, etc.) of astounding variability. This has been verified by two independent studies,[10,11] one of which, an analysis of fifty-four ginseng products, showed that 60 percent of those analyzed were worthless and 25 percent of the sampled products contained no ginseng at all![10]

Of thirty-seven clinical studies published between 1968 and 1990, only fifteen were controlled and eight double-blinded. Seventeen showed improvement in physical performance, eleven improvement in intellectual performance, and thirteen an improvement in mood. Quality of design and statistical analysis in many of the studies is questionable. The German Commission E monograph proposes a daily dosage of 1 to 2 g of the crude drug, or 200 to 600 mg of standardized extracts (calculated to 4 to 7 percent ginsenosides). The drug is used "as a tonic to counteract weakness and fatigue, as a restorative for declining stamina, impaired concentration, and as an aid to convalescence."[12]

Another problem is the herb's relative safety. Much of the concern for toxic side effects stems from a 1979 report by Siegel of a so-called ginseng abuse syndrome in a group of 133 ginseng users.[13] Because of its faulty methodology and lack of even a uniform definition of ginseng, the study has now been thoroughly discredited.[14] Nevertheless, it continues to be cited in the literature with some frequency. Reports attributing estrogenic (female hormone-like) effects to ginseng have been analyzed by Farnsworth. He has concluded that none of them provides any experimental evidence to support such activity.[15] Probably the best documented side effects of ginseng are insomnia, and to a lesser degree, diarrhea, and skin eruptions.[16] Unfortunately, many of the cases reporting other effects have failed to document either the species of ginseng employed or the dose. Even the prolonged or excessive use of ginseng appears to involve relatively low risk.[17]

Another natural drug that is closely enough related to ginseng botanically, chemically, and pharmacologically to be classed as a kind of ginseng is san qui or tienchi ginseng. Also known as sanchi ginseng, it consists of the dried roots of *Panax pseudo-ginseng* Wallich of the family Araliaceae. It is extensively cultivated in Yunnan Province, People's Republic of China. Analyses have shown that tienchi contains 7 to 10.8 percent of crude saponins in comparison to the approximately 4 percent found in normal Asian ginseng. Some of these saponins are identical to the ginsenosides found in the latter species.

In general, tienchi is used as a "tonic," but is recommended in particular for the prevention and treatment of coronary heart disease. No acute or short-term chronic toxicity was noted in small animal tests; clinical studies in 680 cases of coronary disease also showed favorable outcomes.[18] The results of all these Chinese studies require critical evaluation before the safety and efficacy of tienchi can be determined without question.

Eleuthero is not a species of *Panax,* although it is a tall, prickly shrub of the same family, Araliaceae. The part used is the root of *Eleutherococcus senticosus* (Rupr. & Maxim.) Maxim., which is also referred to in the literature as *Acanthopanax senticosus* (Rupr. & Maxim.) Harms. This plant is native to eastern Siberia, Korea, and the Shansi and Hopei Provinces of China. The drug is also known as eleutherococc in the former Soviet Union, as Siberian ginseng in the United States, and as wujiaseng (if intended for sale to America buyers) or ciwujia in China. The common Chinese drug name, ciwujia, has recently been attached to products in the American marketplace, leading the consumer to think it is a new herb. The designation Siberian ginseng is a most unfortunate one because the plant does not even belong to the same genus as the true ginsengs. The name was apparently coined by commercial interests to give an expensive mystique to a relatively cheap drug. Knowledgeable persons prefer the designation eleuthero.

Although the constituents of eleuthero have been designated eleutherosides A through M, not all of these compounds are saponins of the types found in ginseng. Eleutheroside A, for example, is a β-sitosterol glycoside, and eleutheroside B_1 is a coumarin derivative.[19] Still, the same stimulant and tonic effects attributed to ginseng are also claimed for eleuthero. The Chinese report that its adaptogenic or antistress activity is brought about by the combination of contained sterols, coumarins, flavonoids, and polysaccharides.[20]

Large quantities of eleuthero originate in the former Soviet Union, and part of its popularity as a ginseng substitute may derive from its abundance and relatively low cost, at least in comparison to original ginseng. Nevertheless, the same cautions noted for all of these saponin-containing drugs also apply to eleuthero. Lack of standardization of active principles and insufficiently tested clinical effects in human beings speak against their indiscriminate use.[21]

A recent report that consumption of so-called Siberian ginseng by an expectant mother resulted in the birth of an especially hirsute (hairy) infant has stimulated a much closer scrutiny of the herb than ever before. The new information uncovered provides an herbal object lesson far beyond the purported androgenic (male-sex-hormone-like) effect of eleuthero.[22] In the first place, the attending physicians confused the product labeled Siberian ginseng with authentic ginseng derived from *Panax* species.[23] This makes a strong case for the use of the name eleuthero to identify this herb. Secondly, analysis of the product consumed revealed that it was not Siberian ginseng (eleuthero) at all but was the bark of silk vine (*Periploca sepium* Bunge of the family Asclepiadaceae), another herb used in Chinese traditional medicine.[24] Then, because *Periploca* is unlikely to induce androgenic effects, it seemed highly probable that the mislabeled Siberian ginseng (really *Periploca*) was adulterated with something else. Analyses of three different lots of so-called Siberian ginseng revealed that only one was properly identified. However, the one that did contain eleuthero proved to be adulterated by the addition of about 0.5 percent caffeine.[25] Finally, investigation revealed that the powdered eleuthero that is imported into the United States is often misidentified. As a result of several factors, including nomenclatural confusion, it may derive from six or more closely related plants and at least two unrelated species.[26]

A recent case report by a Canadian physician seems to further perpetuate confusion. A seventy-four-year-old man who had been taking digoxin for many years had abnormally elevated serum levels of that drug. Levels remained high even after digoxin therapy was discontinued. The physician then discovered that the patient was taking a "Siberian ginseng" product. After stopping use of the product, serum digoxin levels returned to normal, and digoxin treatment resumed. Several months later, the patient again started taking "Siberian ginseng," and serum digoxin levels rose. Use of "Siberian ginseng" was stopped, and the levels returned to normal. The abnormally high levels of digoxin were attributed to "Siberian ginseng." Unfortunately, the plant material was not analyzed before or after the report was published. It appears this may have been another case of confusion between Siberian ginseng and *Periploca sepium. Periploca sepium* does contain glycosides that may have been responsible for potentiating the effect of digitalis glycosides. These cases highlight the need for proper botanical identification of herbal products, both in the marketplace and in scientific publications.[27,28]

All of this is a sad commentary on the quality of some herbal products currently available in the marketplace. Because standards of quality are lacking, the only recourse the consumer has at present is to obtain herbs or herbal products only from individuals or organizations with the highest possible ethical standards and reputations.

Four clinical studies financed by a single manufacturer and published in the period 1994 to 1996 have provided evidence that a standardized extract of ginseng (G115), in combination with diethylaminoethanol bitartrate, multivitamins, and minerals, promoted an improved quality of life, a feeling of well-being, and a beneficial effect on functional fatigue in comparison to placebos.[29,30,31,32] One of the studies utilizing the additives without the ginseng extract as a placebo noted a significant improvement in quality of life of the subjects as measured by a standardized eleven-item questionnaire.[32] Therefore, it is likely that at least some of the beneficial effects are attributable to the ginseng.

As a result of these recent studies, I am much less skeptical than formerly about the utility of ginseng in humans. Still, until additional well-designed clinical trials on ginseng without additives are conducted, it will not be possible to determine the activity of the herb with certainty. The many other companies, as well as certain governments, that realize a substantial financial return from the sale of this high-priced botanical should devote some of their profits toward this worthwhile objective. Such studies are long overdue.

REFERENCES

1. *Abstracts of Korean Ginseng Studies (1687-1975).* The Research Institute, Office of Monopoly, Seoul, Republic of Korea, 1975.

2. A. O. Tucker, J. A. Duke, and S. Foster. Botanical Nomenclature of Medicinal Plants. In *Herbs, Spices, and Medicinal Plants: Recent Advances in Botany, Horticulture, and Pharmacology,* Volume 4, L. E. Craker and J. E. Simon (Eds.). Oryx Press, Phoenix, Arizona, 1989, pp. 169-242.

3. C. S. Robbins. *American Ginseng: The Root of North America's Herb Trade.* TRAFFIC North America, World Wildlife Fund, Washington, DC, 1998.

4. N. R. Farnsworth. *Tile & Till* 59: 30-32, 1973.

5. J. P. Hou. *The Myth and Truth About Ginseng.* A. S. Barnes and Co., South Brunswick, New Jersey, 1978.

6. N. R. Farnsworth, op. cit.

7. V. E. Tyler, L. R. Brady, and J. E. Robbers. *Pharmacognosy,* Ninth Edition. Lea and Febiger, Philadelphia, Pennsylvania, 1988, pp. 473-475.

8. A. Y. Leung and S. Foster. *Encyclopedia of Common Natural Ingredients Used in Food, Drugs, and Cosmetics,* Second Edition. John Wiley and Sons, New York, 1996, pp. 277-281.

9. J. P. Hou, op. cit.

10. L. E. Liberti and A. Der Marderosian. *Journal of Pharmaceutical Sciences* 67: 1487-1489, 1978.

11. W. Ziglar. *Whole Foods* 2(4): 48-53, 1979.

12. V. Schulz, R. Hänsel, and V. E. Tyler. *Rational Phythotherapy: A Physicians' Guide to Herbal Medicine,* Third Edition. Springer, Berlin, 1998, pp. 270-272.

13. R. K. Siegel. *Journal of the American Medical Association* 241: 1614-1615, 1979; 243: 32, 1980.

14. M. Castleman. *Herb Quarterly* 48: 17-24, 1990.

15. N. R. Farnsworth. Personal communication, January 29, 1991.

16. C. A. Baldwin, L. A. Anderson, and J. D. Phillipson. *Pharmaceutical Journal* 237: 583-586, 1986.

17. R. F. Chandler. *Canadian Pharmaceutical Journal* 121: 36-38, 1988.

18. Tienchi-Ginseng. China National Native Produce & Animal By-Products Import & Export Corp., Yunnan Native Produce Branch, Kunming, 1979, not pgd.

19. J. Connert: *Deutsche Apotheker Zeitung* 120: 735-736, 1980.

20. Wujiaseng. China National Native Produce & Animal By-Products Import & Export Corp., Heilungkiang Native Produce Branch, nd, not pgd.

21. V. E. Tyler, L. R. Brady, and J. E. Robbers, op. cit., pp. 473-475.

22. G. Koren, S. Randor, and D. Dannema. *Journal of the American Medical Association* 264: 2866, 1990.

23. D. V. C. Awang. *Journal of the American Medical Association* 265: 1828, 1991.

24. D. V. C. Awang. Personal communication, April 18, 1991.

25. Ibid., May 21, 1991.

26. A. Y. Leung. Unpublished report, June 27, 1991.

27. S. MacRae. *Canadian Medical Association Journal* 155: 292-295, 1996.

28. S. Foster. *Siberian Ginseng,* Eleutherococcus senticosus, *Botanical Series No. 30,* Second Edition. American Botanical Council, Austin, Texas, 1996, 8 pp.

29. I. Wiklund, J. Karlberg, and B. Lund. *Current Therapeutic Research* 55: 32-42, 1994.

30. M. Neri, E. Andermarcher, J. M. Pradelli, and G. Salvioli. *Archives of Gerontology and Geriatrics* 21: 241-252, 1995.

31. M. LeGal, P. Cathebras, and K. Strüby. *Phytotherapy Research* 10: 49-53, 1996.

32. A. Caso Marasco, R. Vargas Ruiz, A. Salas Villagomez, and C. Begoña Infante. *Drugs in Experimental Clinical Research* 22: 323-329, 1996.

Goldenseal

Goldenseal or hydrastis is a native American drug, having been introduced to the early settlers by the Cherokee Indians, who used it primarily for skin diseases and as a wash for sore eyes. It consists of the rhizome and roots of the small forest plants *Hydrastis canadensis* L., a member of the family Ranunculaceae.[1] In recent years, increased demand has resulted in supply shortages and apparent scarcity in its native haunts, prompting it to be nominated for a CITES Appendix II listing, which went into effect in September 1997. CITES (Convention on International Trade in Endangered Species of Wild Fauna and Flora) is an international treaty with over 160 signatory nations. The treaty has established a permit system to regulate trade in endangered species or species at serious risk. Species banned from international trade (such as leopard skins) are placed in Appendix I. Appendix II listings, such as that for American ginseng, and now goldenseal, control and monitor international trade, "in order to avoid utilization incompatible with their survival." The U.S. Fish and Wildlife Service, responsible for CITES implementation, must now develop a system to monitor goldenseal in international trade.

Over the years, goldenseal acquired a considerable reputation as a general bitter tonic and as a remedy for various gastric and genitourinary disorders. The drug was a prominent ingredient in many turn-of-the-century proprietary medicines, including Dr. Pierce's Golden Medical Discovery:

> A few of my symptoms were: Heartburn and fullness after eating, sometimes pain in my bowels, headache, poor appetite, and bad taste in my mouth. At night I was feverish, with hot flushes over my skin. After taking Dr. Pierce's Golden Medical Discovery I was relieved of all these symptoms, and I feel perfectly well today.
>
> J. P. McAdams
> Elon College, North Carolina[2]

Whatever activity is possessed by goldenseal can be attributed to its contained alkaloids, especially hydrastine and berberine. The latter is also responsible for the drug's characteristic golden color. Although these alkaloids, administered by injection, do exert some minor actions on circulation, uterine tone and contractility, and the central nervous system, unless extremely large, near-toxic doses are administered, the effects are too uncertain to be therapeutically useful. The alkaloids are not absorbed following oral administration so, when taken by mouth, they, and the herb containing them, produce no systemic effects. Goldenseal is no longer even discussed in modern works on pharmacology, and Sollmann has concluded that it "has few, if any, rational indications."[3] Nevertheless, it continues to occupy a place of prominence in modern herbals; one of them devotes seventy-eight pages to a discussion of the various aspects of this interesting drug.[4]

A survey of folk medicine in Indiana produced so many endorsements of goldenseal tea as a useful rinse to relieve sore mouth, cracked and bleeding lips, canker sores, and related problems that this use must at least be mentioned here.[5] The previously mentioned alkaloids do possess both astringent and weak antiseptic properties on local application, which cause goldenseal tea to be a modestly effective topical treatment in many such cases.

When administered by injection, berberine has been shown to stimulate digestion and the secretion of bile, lower blood pressure, and produce a vasoconstricting effect and an adrenolytic effect (inhibiting the action of adrenaline). Most pharmacological actions have been attributed to the effects of a single alkaloid isolated from the root. Italian researchers recently found that the individually isolated alkaloids produced a dose-related inhibition of contraction responses when exposed to low doses of adrenaline. At higher adrenaline concentrations, none of the isolated alkaloids inhibited contractions induced by adrenaline. This was true for the alkaloids berberine, canadine, and canadaline. Hydrastine was found to be completely inactive. However, the total extract of the root inhibited adrenaline-induced contractions at both low and high dosage levels. A mixture of berberine, canadine, and canadaline (with hydrastine excluded) had an even more potent adrenaline-inhibiting effect than the total extract of goldenseal root, suggesting that the sum of the combined alkaloids is greater than corresponding amounts of the isolated alkaloids.[6]

Several years ago, goldenseal gained considerable publicity from the claim that usually as an herbal tea, it prevented the detection of morphine in urine specimens following heroin use. It thus achieved considerable popularity among certain heroin addicts who were patients in methadone or similar drug rehabilitation programs. Scientific studies have revealed no basis for this claim. Goldenseal neither prevents morphine detection nor does it "flush" that compound from the body.[7] Nevertheless, this myth continues to exist in expanded form today. Personnel in health food stores and head shops (stores specializing in drug-abuse paraphernalia) on the West Coast are reported to be recommending the consumption of goldenseal as a means of thwarting the now widely used urine tests designed to detect illegal use of marijuana and/or cocaine. Authorities, however, disagree.[8] As *High Times* has repeatedly warned, *no* edible substance, taken internally, will have any effect at all on drug-urinalysis machines (except to possibly promote false-positive readings). Foster has discussed the fictional origins of the myth that goldenseal affects drug testing, tracing it back to pharmacist John Uri Lloyd's 1900 novel *Stringtown on the Pike*.[9]

REFERENCES

1. S. Foster. *Goldenseal,* Hydrastis canadensis, *Botanical Series No. 309,* Second Edition. American Botanical Council, Austin, Texas, 1996, 8 pp.

2. R. V. Pierce. *The People's Common Sense Medical Adviser,* Sixty-Second Edition. World's Dispensary Printing Office and Bindery, Buffalo, New York, 1895, p. 589.

3. T. Sollmann. *A Manual of Pharmacology,* Seventh Edition. W. B. Saunders, Philadelphia, Pennsylvania, 1948, pp. 257-258.

4. L. Veninga and B. R. Zaricor. *Goldenseal/Etc.* Ruka Publications, Santa Cruz, California, 1976, pp. 1-78.

5. V. E. Tyler. *Hoosier Home Remedies.* Purdue University Press, West Lafayette, Indiana, 1985, pp. 138-139.

6. M. Palmary, M. F. Cometa, and M. G. Leone. *Phytotherapy Research* 10: S47-S49, 1996.

7. J. A. Ostrenga and D. Perry. *PharmChem Newsletter* 4(1): 1-3, 1975.

8. M. W. Montague. *High Times* 135: 15, 85, 93, 1986.

9. S. Foster. *HerbalGram* 21: 7, 35, 1989.

Gotu Kola

Two leaves a day will keep old age away.

Ancient Sinhalese Proverb

The leaves in this saying are those of *Centella asiatica* (L.) Urb. (*Hydrocotyle asiatica* L.) of the family Apiaceae, a slender creeping plant that is especially abundant in the swampy areas of India and Sri Lanka, South Africa, and the tropical regions of the New World. It is commonly called gotu kola but is also known as hydrocotyle or Indian pennywort.

In Sri Lanka, it was observed that elephants, noted for their longevity among beasts, fed extensively on the plant. This gave rise to the reputation of the herb as a longevity promoter for people. Eating a few leaves daily was thought to "strengthen and revitalize worn out bodies and brains." Gotu kola has also been recommended as a treatment for mental troubles, high blood pressure, abscesses, rheumatism, fever, ulcers, leprosy, skin eruptions, nervous disorders, and jaundice.[1] More recently, the herb has acquired a considerable reputation as an aphrodisiac, an agent that stimulates sexual desire and ability.[2] The crushed leaves are commonly consumed by people in Sri Lanka, either in the form of a salad or as a hot beverage.

Scientific studies have shown that in relatively large doses the drug has a definite sedative effect in small animals. This activity comes from two saponin glycosides, designated brahmoside and brahminoside.[3] Another glycoside, known as madecassoside, exhibits some anti-inflammatory activity, and still another, asiaticoside, apparently stimulates wound healing.[4] However, there is currently no evidence to support the use of gotu kola as a longevity promoter or to substantiate any of the other extravagant claims for its use as a revitalizing and healing herb. Substantive data on its safety and efficacy are simply nonexistent.

In Vietnam, it is considered an edible weed. Foster, in a recent trip to Vietnam's central highlands, observed it growing in ginseng beds and suggested to his hosts that it might be a potential export crop. The Vietnamese reacted in a manner similar to that of a Kansas farmer to whom someone suggested that he stop growing wheat and plant dandelion instead. "It is a weed!" they exclaimed.

A word of caution is necessary about the name of this herb. The similarity in spelling has caused gotu kola to be confused with the dried cotyledon (seed leaf) of *Cola nitida* (Vent.) Schott & Endl., otherwise known as kolanuts, kola, or cola.[5] This plant material, well-known as an ingredient of Coca-Cola, contains up to 3.5 percent caffeine, a constituent not present in gotu kola. Whatever else it may be, gotu kola is certainly not a stimulant.

REFERENCES

1. V. De Silva. *Medical Herbalist* 11: 159, 1936.

2. A. Gottlieb. *Sex Drugs and Aphrodisiacs.* High Times/Level Press, New York and San Francisco, 1974, pp. 37-38.

3. A. S. Ramaswamy, S. M. Periyasamy, and N. Basu. *Journal of Research in Indian Medicine* 4: 160-175, 1970.

4. P. H. List and L. Hörhammer (Eds.). *Hagers Handbuch der Pharmazeutischen Praxis,* Fourth Edition, Volume 3. Springer-Verlag, Berlin, 1972, pp. 792-793.

5. R. K. Siegel. *Journal of the American Medical Association* 236: 473-476, 1976.

Grape Seed Extract

The glory of grapes was once relegated to enjoying a glass of fine wine. Today, *Vitis vinifera* L., a perennial woody vine of the family Vitaceae, native to Asia minor, is becoming known as a source of a dietary supplement, exclusive of the health benefits attributed to red wine. Agronomists thought they were doing consumers a favor when they developed seedless grapes, but as it turns out, seed extracts may be the most valuable part of the grape, at least on a weight-to-dollar ratio. Grapes, and their dried counterparts, raisins, are well-known for nutritional benefits, particularly as a source of iron. They are among the oldest cultivated fruits, thought to have originated in the Caspian Sea region of Western Asia. Grapes (and wine) are frequently mentioned in the Bible. They were grown in Egypt at least 6,000 years ago.[1] It is estimated that grapes were introduced to Greece between the eighteenth and sixteenth centuries B.C. The common cultivated grape is represented by more than 8,000 named varieties, about 20 percent of which are still grown today. The medicinal value of grapes (often in the form of wine) was also praised by such luminaries of the ancient world as Theophrastus, Dioscorides, Pliny, and Galen.[2]

As one of the most important economic plants, the chemistry of grapes has been extensively studied. Interest in grape seeds stems from oligomers or polymers of catechin and epicatechin present in the skin, particularly of red grapes, but even more abundant in the seeds. These compounds, known a procyanidins (also called condensed tannins, pycnogenols, leucoanthocyanins, oligomeric proanthocyanidins or OPCs), are concentrated in proprietary products now used in Europe and America for the treatment of microcirculatory disorders of various origins. Standardized by a proprietary process, grape seed polyphenol extracts contain procyanidin dimers, trimers, tetramers, oligomers, along with catechin and eipcatechin. The seeds also contain a fixed oil with various essential fatty acids and tocopherols.[3]

Antioxidant activity, including scavenging of free radicals and inhibition of lipid peroxidation, by grape seed extracts or OPCs has been demonstrated in various experimental models. Topically, procyanidins have been shown to prevent degradation of structural components of extravascular matrix such as collagen, elastin, and hyaluronic acid. In the endothelium, the free-radical scavenging and antiperioxidative effects of procyanidins have been shown to contribute to structural integrity by inhibiting proteases such as collagenase, elastase, hyaluronidase, and beta-glucuronidase. Grape seed procyanidins have been shown to be effective in reducing capillary permeability in several animal models. At least four controlled clinical studies have evaluated grape seed extracts in the treatment of peripheral venous insufficiency, with cautiously positive results when compared with placebo or conventional drugs, such as improvements related to pain, nocturnal cramps, edema, and paresthesias. The studies involved small numbers of patients (less than 100). Five clinical studies have evaluated the potential of the extract in various opthalmological pathologies. They measured ocular stress produced by computer screens or myopia, macular degeneration, and reducing glare from bright lights, with promising results.[4,5]

Much of the scientific work on grape seed extracts has been conducted in France and published in the French scientific literature, somewhat limiting access to peer review. Italian studies, generally published in English, are more accessible.[2] Grape seed products have been marketed in France for decades, primarily for the treatment of peripheral venous insufficiency, as well as an ingredient in numerous cosmetic products promising a reduction in the signs of aging. Less research has been conducted on pine bark extracts, also concentrating procyanidins, though similar claims are made of pine bark products. To confuse matters more, use of the term "pycnogenol," first coined in 1979 by J. A. Masquelier, as a generic term to describe proanthocyanidins has been trademarked by different corporate entities in different countries, leading to confusion. In the American market Pycnogenol is a registered trademark of a pine bark product derived from *Pinus pinaster* Ait. (usually referenced by the obsolete name *Pinus maritima* Mill.).

French researchers turned to the abundant by-product of the Bordeaux region's wine industry, grape seeds, only after the original raw material—peanut skins—for the first OPC-based products dried up in France. African peanut growers began supplying France with shelled rather than whole peanuts in 1950.[6] Despite the abundance of this raw material in the United States, it is doubtful that peanut skin extracts will replace grape seed extracts in the American market. Much of the quality grape seed extract now marketed in the United States is obtained from Indena, a producer in Italy.

REFERENCES

1. S. Foster. *Health Foods Business* April: 42-43, 1997.
2. E. Bombardelli and P. Morazzoni. *Fitoterapia* 66(4): 291-371, 1995.
3. Ibid.
4. Ibid.
5. *Lawrence Review of Natural Products,* September 1995.
6. J. A. Masquelier. *A Lifetime Devoted to OPC and Pycnogenols.* Alfa Omega Editrice, Rome, Italy, 1997.

Hawthorn

Although some of the medicinal properties of the fruits or haws of this small to medium-sized tree, *Crataegus laevigata* (Poir.) DC. (*C. oxycantha* auct., non L.) of the family Rosaceae, were known to Dioscorides in the first century A.D., and the drug continued to be mentioned by Gerard and other early herbalists, hawthorn was not widely used therapeutically until very recent times.[1] Even now, the drug, which may also be obtained from *Crataegus monogyna* Jacq., is little known in the United States. But hawthorn flowers, fruits, and leaves are prominent in Continental medicine. About three dozen different preparations containing extracts of these plant parts, either singly or in combination with other drugs, are currently marketed in Germany. That figure does not include a number of herbal mixtures or teas that have hawthorn as one of their major ingredients.

However, this interesting plant has not been entirely neglected, even in this country. A 125-page book titled *The Hawthorn Berry for the Heart,* devoted to a discussion of hawthorn's medicinal properties, was published in 1971.[2] Hawthorn is described in it and most other modern herbals as a valuable drug for the treatment of various heart ailments and circulatory disorders.[3,4,5] Wrongly believing that only the bark contained cardioactive principles, Gibbons experimented with haw jelly and haw marmalade, even giving some to his friends at Christmas.[6] He described these mildly astringent products as a very pleasant way of treating a sore throat. Since the haws, leaves, and flowers all contain compounds that do affect the heart and circulatory system, products containing them should not be consumed indiscriminately. Interestingly, it was not until 1974, some years after Gibbons made his comments, that a scientific study revealed that the bark of a *Crataegus* species contained any active principles.[7]

Other 1970s' research revealed some interesting properties of hawthorn. It acts on the body in two ways. First, it dilates the blood vessels, especially the coronary vessels, reducing peripheral resistance and thus lowering the blood pressure. It is thought to reduce the tendency to experience angina attacks. Second, it apparently has a direct, favorable effect on the heart itself, which is especially noticeable in cases of heart damage. Hawthorn's action is not immediate, but rather develops very slowly. Its toxicity is low as well, becoming evident only in large doses. It therefore seems to be a relatively harmless, mild heart tonic that apparently yields good results in many conditions for which this kind of therapy is required.[8]

The active principles of hawthorn are probably a mixture of pigments known as flavonoids, large numbers of which are contained in the various plant parts. So-called oligomeric procyanidins (dehydrocatechins) seem to be particularly active.[9] They also produce marked sedative effects that indicate an action on the central nervous system.

In recent years, the German Commission E has more narrowly defined hawthorn preparations deemed to possess therapeutic value. A revised monograph, published in 1994, recognizes a fixed combination of the flowers, leaves, and fruits, plus a preparation of the leaves and flowers. Both are water-and-alcohol extracts with an herb-to-extract ratio of 5-7:1. These preparations are calculated to deliver 4 to 20 mg of flavonoids (calculated as hyperoside) and 30 to 160 mg of oligomeric procyanidins (calculated as epicatechin) in a daily dose of 160 to 900 mg of hawthorn extract. Dosing is determined by the patient's physician. These oral dosage forms are continued for at least six weeks. Other preparations, including alcoholic extract of the leaves or flowers alone, may be useful, but clinical studies are lacking. Preparations containing hawthorn leaf, berry, or flowers alone in mono preparations are not allowed to carry therapeutic claims since effectiveness has not been documented.[10]

Further scientific studies may eventually substantiate these findings; in fact, studies are urgently needed for a drug as potentially valuable as this one.[11,12] Until additional research has been carried out, prospective users of hawthorn for heart and circulation problems should consider all the consequences. Users of self-selected medicines almost always do so as a result of self-diagnosis. This is a very dangerous practice when such vital systems of the human body as the heart and blood vessels are involved. For this reason, self-treatment with hawthorn is neither advocated nor condoned.

REFERENCES

1. E. Steinegger and R. Hänsel. *Lehrbuch der Pharmakognosie,* Third Edition. Springer-Verlag, Berlin, 1972, pp. 146-147.

2. J. I. Rodale. *The Hawthorn Berry for the Heart.* Rodal Books, Emmaus, Pennsylvania, 1971.

3. D. Law. *The Concise Herbal Encyclopedia.* Saint Martin's Press, New York, 1973, pp. 56-57, 116.

4. R. Lucas. *Nature's Medicines.* Wilshire Book Co., North Hollywood, California, 1977, pp. 188-189.

5. M. Tierra. *The Way of Herbs.* Unity Press, Santa Cruz, California, 1980, p. 97.

6. E. Gibbons. *Stalking the Healthful Herbs,* Field Guide Edition. David McKay Co., New York, 1970, pp. 171-174.

7. E. B. Thompson, G. H. Aynilian, P. Gora, and N. R. Farnsworth. *Journal of Pharmaceutical Sciences* 63: 1936-1937, 1974.

8. E. Steinegger and R. Hänsel, op. cit., pp. 146-147.

9. W. Rewerski, T. Piechocki, M. Rylski, and S. Lewak. *Arzneimittelung* 21: 886-888, 1971.

10. V. Schulz, R. Hänsel, and V. E. Tyler. *Rational Phytotherapy: A Physicians' Guide to Herbal Medicine,* Third Edition. Springer, Berlin, 1998, pp. 89-99.

11. N. W. Hamon. *Canadian Pharmaceutical Journal* 121: 708-709, 724, 1988.

12. C. Hobbs and S. Foster. *HerbalGram* 22: 19-33, 1990.

Hibiscus

A variety of names—hibiscus, roselle, Sudanese tea, red tea, and Jamaica sorrel—designate the flowers (actually calyces and bracts) of *Hibiscus sabdariffa* L. This red-flowered annual herb of the family Malvaceae is widely cultivated throughout the tropics, reaching a height of four to five feet or more. Its flower heads are collected when immature and are highly prized for making jams, jellies, sauces, and acid beverages.[1] The floral parts make a pleasant tea and are used by themselves or mixed with other herb teas.[2]

Hibiscus contains various anthocyanins and other pigments plus relatively large amounts of oxalic, malic, citric (12 percent to 17 percent), and tartaric acid, as well as up to 28 percent of hibiscic acid (the lactone of a hydroxycitric acid).[3] These plant acids are responsible for the tart, refreshing taste of various hibiscus beverages and foods. They probably also account for the mild laxative and diuretic effects attributed to the plant. Appreciable quantities of water-soluble mucilaginous polysaccharides are also found in the herb. It is believed that they may be responsible for some of the numerous physiological effects postulated for hibiscus.[4,5]

Scientific verification of the various pharmacological properties is required. For example, its hypotensive (blood-pressure lowering) action was observed only on direct injection of an extract into the vein of a dog; the effect was very brief, even when large amounts were given.[6] Critical reevaluation of the experiments in which hibiscus extracts were found to inhibit the growth of the tubercle bacillus is also advisable.[7] The recent contradictory report of both stimulatory and inhibitory effects of hibiscus on various muscle preparations in vitro must also be considered as merely a preliminary observation.[8] Yet another preliminary investigation found that protocatechuic acid isolated from the plant had an inhibitory effect on tumor promotion in mouse skin.[9] Protection of oxidative stress in rat primary hepatocytes was recently reported by a Taiwanese research group.[10] These reports present further research leads rather than reproduced, confirmed activities.

Widely employed throughout the world as a beverage and food flavor, hibiscus imparts a taste that is obviously pleasant to many people. So there appears to be no reason to discourage anyone from using it for this purpose.

REFERENCES

1. L. H. Bailey and E. Z. Bailey. *Hortus Third.* Macmillan, New York, 1976, pp. 562, 982-983.

2. M. Stuart (Ed). *The Encyclopedia of Herbs and Herbalism.* Grosset and Dunlap, New York, 1979, p. 201.

3. A. Y. Leung and S. Foster. *Encyclopedia of Common Natural Ingredients Used in Food, Drugs, and Cosmetics,* Second Edition. John Wiley & Sons, New York, 1996, pp. 444-446.

4. M. Franz and G. Franz. *Zeitschrift für Phytotherapie* 9: 63-66, 1988.

5. B. M. Müller and G. Franz. *Deutsche Apotheker Zeitung* 130: 329-333, 1990.

6. A. Sharaf. *Planta Medica* 10: 48-52, 1962.

7. A. Sharaf and A. Gineidi. *Planta Medica* 109-112, 1963.

8. M. B. Ali, W. M. Salih, A. H. Mohamed, and A. M. Homeida. *Journal of Ethnopharmacology* 31: 249-257, 1991.

9. T. H. Tseng, J. D. Hsu, M. H. Lo, C. Y. Chu, F. P. Chou, C. L. Huang, and C. J. Wang. *Cancer Letters* 24: 199-207, 1998.

10. T. H. Tseng, E. S. Kao, C. Y. Chu, F. P. Chou, H. W. Lin Wu, and C. J. Wang. *Food and Chemical Toxicology* 35: 1159-1164, 1997.

Honey

Honey is the saccharine secretion deposited in the honeycomb by the bee, *Apis mellifera L.,* a well-known insect of the family Apidae. To prepare this product, the worker bee collects nectar from various flowers and stores it briefly in its crop or honey bag. There it is acted upon by secretions from glands in the bee's head and thorax that bring about various changes, particularly conversion of much of the contained sucrose (cane sugar) to so-called invert or simple sugars. On returning to the hive, the insect regurgitates the viscous liquid, now known as honey, into the wax comb where it is extracted for marketing.[1]

The 1811 edition of *The Edinburgh New Dispensatory* informs us that, "From the earliest ages, honey has been employed as a medicine . . . it forms an excellent gargle and facilitates the expectoration of viscid phlegm; and it is sometimes employed as an emollient application to abscesses, and as a detergent to ulcers."[2] More recent advocates of the medicinal use of honey have greatly expanded its purported virtues. D. C. Jarvis, a Vermont physician who advocated a mixture of honey and vinegar as a cure-all, wrote a detailed account of the so-called therapeutic uses of honey. He claimed that it improved digestion; attracted fluid and thereby facilitated the healing of wounds and ulcers; helped the body destroy harmful germs; was an excellent food supplement because of its content of vitamins, minerals, and enzymes; was a useful laxative; had a sedative effect; and helped to relieve arthritis pain. As if these seven actions were not sufficient, he also maintained that persons who ate honey and kept bees were entirely free from cancer and paralysis.[3]

Because of its commercial importance as a nutrient and sweetener, honey has been subjected to extensive chemical analyses. It consists of about 40 percent fructose (fruit sugar), 35 percent glucose (grape sugar), 4 percent other sugars including sucrose, 18 percent water, and 3 percent other substances such as aromatic principles and tannin. Vitamins, minerals, and proteins (enzymes) are also present but in such tiny amounts as to preclude any therapeutic utility or, for that matter, any real nutritional significance.[4]

Of course, honey is a tasty and useful sweetening agent; it serves as a rapid source of energy because it contains simple sugars. Honey also is still used in folk medicine for its demulcent or soothing effects, particularly in various cough remedies. However, there is no evidence to support claims of any sedative action or that it will relieve the pain of arthritis or any other affliction. References to its role in preventing or curing cancer or paralysis are gross exaggerations. Whatever antibacterial properties honey may possess are due primarily to its high sugar content. Once diluted by contact with body or other fluids, any such effect is lost.

Honey often contains spores of *Clostridium botulinum*, the organism responsible for infant botulism. It is believed that in older children and adults these spores germinate without producing ill effects. However, in some infants they may cause serious illness and even death. In 1976, for example, forty-three cases of infant botulism were diagnosed in California. Of those, thirteen were attributed to *Clostridium botulinum* in honey.[5] For this reason, it has been recommended that honey not be given to infants under the age of one year.[6]

Like the queen in the Mother Goose rhyme, I do savor the taste of honey:

> The queen was in the parlor,
> Eating bread and honey.

It is an elegant spread for bread or rolls. Honey added to hot lemonade effectively soothes a sore throat and may, in one way or another, even help ease the cold that caused it. Beyond this, it is unrealistic to expect any significant medicinal value from the product. Your expectations simply will not be realized.

REFERENCES

1. H. W. Felter and J. U. Lloyd. *King's American Dispensatory,* Eighteenth Edition, Volume 2. The Ohio Valley Co., Cincinnati, Ohio, 1900, pp. 1247-1249.

2. A. Dunca Jr. *The Edinburgh New Dispensatory,* Sixth Edition. Bell and Bradfute, Edinburgh, Scotland, 1811, pp. 324-325.

3. D. C. Jarvis. *Arthritis and Folk Medicine,* Reprint Edition. Fawcett Publications, Greenwich, Connecticut, 1960, pp. 134-137.

4. D. Johnson. In *Whole Foods Natural Foods Guide.* And/Or Press, Berkeley, California, 1979, pp. 76-79.

5. A. Y. Leung and S. Foster. *Encyclopedia of Common Natural Ingredients Used in Food, Drugs, and Cosmetics*, Second Edition. John Wiley and Sons, New York, 1996, pp. 299-300.

6. R. A. Mangione. *American Pharmacy* NS23: 5, 1983.

Hops

The hop plant, *Humulus lupulus* L., a member of the family Cannabidaceae, is a perennial climbing vine that bears scaly cone-like fruits known as hops. These fruits, technically called strobiles, are covered with glandular hairs containing resinous bitter principles that account for the use of hops in brewing and in medicine. Hops are extensively cultivated in the Czech Republic, England, Germany, the United States, South America, and Australia. They are collected in September when ripe and marketed after careful drying.[1]

Although hops have been used in beer for over 1,000 years, primarily for their bitter taste and preservative action, their medicinal or tonic properties were apparently also valued from very early times. It was observed that hop pickers tired easily, apparently as a result of the accidental transfer of some hop resin from their hands to their mouths, and the drug gained a reputation as a sedative. Pillows filled with hops have been used for sleeplessness and nervous conditions. A small bag of hops, wetted with alcohol and placed hot on the afflicted area, was said to reduce local inflammation. Aqueous extracts made with boiling water have been used as tonics.

"TAKE HOP BITTERS three times a day, and you will have no doctor bills to pay," was the advertising slogan of one patent medicine manufacturer in the last century. In addition to hops, this popular product contained some buchu, mandrake, dandelion, and 30 percent alcohol. During the four-year period preceding 1884, $2.75 million worth of this nostrum was sold.[2]

Chemically unstable polyphenolic principles, especially humulone and lupulone, are present in the resin of hops. They, or closely related conversion products, are responsible for the plant's bitter and bacteriostatic properties. Unfortunately, the content of these compounds varies appreciably in different varieties of hops, and in addition, these compounds are quite unstable in the presence of air and light. One study has shown that after nine months' storage, hops retained only about 15 percent of their original activity.[3]

Early studies on hops failed to identify specific sedative principles, and their value, particularly when used in the form of a pillow, was thought to be more magical than medicinal. More recently, a volatile alcohol, 2-methyl-3-butene-2-ol (dimethylvinyl carbinol), has been isolated from hops and is believed to account for at least part of the plant's sedative properties.[4] Present in fresh hops in very small amounts, the concentration of the alcohol increases on drying to reach a maximum value of about 0.15 percent within a two-year period.

Mobility tests in rats verified the sedative-hypnotic activity of the alcohol, and pharmacologically active concentrations of it were detected in freshly prepared hop teas. Although studies thus far carried out do not provide an explanation for all of the salutary effects attributed to hops by folklore, they do supply, for the first time, a logical scientific basis for at least part of their tranquilizing action. Continuing investigations will probably eventually supply the rest of the story regarding their benefit to humankind. Based on a reasonable certainty of efficacy and safety, rather than current controlled clinical studies, in Germany, a dose of 0.5 g of dried hops or extract equivalents is allowed to be labeled "for discomfort due to restlessness or anxiety and sleep disturbances."[5]

Hops are closely related botanically to marijuana, and some writers advocate smoking the plant material to obtain a mild euphoria.[6] This practice cannot be recommended since unpleasant side effects are common, and the safety of smoking hops remains in doubt.

REFERENCES

1. V. E. Tyler, L. R. Brady, and J. E. Robbers. *Pharmacognosy,* Ninth Edition. Lea and Febiger, Philadelphia, Pennsylvania, 1988, p. 477.

2. H. W. Holcombe. *Patent Medicine Tax Stamps.* Quarterman Publications, Lawrence, Massachusetts, 1979, pp. 248-251.

3. G. Berndt. *Deutsche Apotheker Zeitung* 106: 158-159, 1966.

4. R. Wohlfart. *Deutsche Apotheker Zeitung* 123: 1637-1638, 1983.

5. V. Schulz, R. Hänsel, and V. E. Tyler. *Rational Phytotherapy: A Physicians' Guide to Herbal Medicine*, Third Edition. Springer, Berlin, 1998, pp. 81-83.

6. L. A. Young, L. G. Young, M. M. Klein, D. M. Klein, and D. Beyer. *Recreational Drugs.* Collier Books, New York, 1977, p. 98.

Horehound

Horehound or hoarhound, is a hairy, bitter-aromatic, perennial herb native to the Mediterranean region of Europe and Asia but naturalized in North America. Designated botanically as *Marrubium vulgare* L. and classified in the mint family, or Lamiaceae, its leaves and flowering tops have long been widely used, both as a folk medicine and a flavoring agent. The common name, horehound, derives from the abundant whitish hairs that cover the plant's leaves, giving them a hoary appearance; hound refers to the use of the plant by the ancient Greeks to treat mad-dog bite.

Nearly four centuries ago, Gerard praised horehound's usefulness in treating coughs and consumption (tuberculosis), and more recently, Grieve noted its expectorant, tonic, and antiasthmatic properties.[1] It is also said to act as a stomachic, a vermifuge, and in large doses, a purgative. Applied externally, it is recommended as a treatment for wounds. Probably the most common use of horehound today is in lozenges and syrups intended for the relief of coughs and minor throat irritations. Most children today are familiar with the taste of horehound candy.

References attribute the activity of the herb to a combination of marrubiin—a bitter diterpenoid lactone principle which has been isolated from it in concentrations ranging from 0.3 to 1.0 percent—volatile oil, and tannin. However, marrubiin is now known to be an artifact that does not preexist in the plant but is formed from the closely related premarrubiin during the isolation procedure.[2]

Horehound is apparently an effective expectorant. This property is a result, not so much of the contained volatile oil, but of the marrubiin (premarrubiin) that stimulates the secretions of the bronchial mucosa.[3] Other pharmacological effects attributed to the drug are more problematic and require extensive investigation before they can be judged either significant or useful. Marrubiin is reputed to be effective in normalizing cardiac arrhythmias (heart irregularities)—but in large doses can cause them. Marrubic acid, which can be formed from marrubiin, has been shown to stimulate the flow of bile in rats.[4] This might account for the purgative properties observed with large doses, but it is really difficult to evaluate since the marrubic acid content of horehound is unknown. An alcoholic extract has been found to reduce spasms of smooth muscles of the gastrointestinal tract, which might also contribute to this effect.[5]

The pleasant fragrance and taste of horehound volatile oil plus the expectorant action of its bitter principle account for the widespread use of the herb and its extracts for coughs and colds. German health authorities have approved horehound not only for the treatment of such conditions but also as an appetite stimulant and a digestive aid.[6] In contrast, the FDA in this country has conceded the safety of the herb as a food additive but declared it to be an ineffective cough suppressant and remains unconvinced of its effectiveness as an expectorant. It is suggested that interested consumers try a little horehound candy the next time they have a cough and determine for themselves which judgment regarding the utility of the herb is correct.

REFERENCES

1. M. Grieve. *A Modern Herbal,* Volume 1. Dover Publications, New York, 1971, pp. 415-416.
2. M. S. Henderson and R. McCrindle. *Journal of the Chemical Society*, Section C: 2014-2015, 1969.
3. P. H. List and L. Hörhammer (Eds.). *Hagers Handbuch der Pharmazeutischen Praxis,* Fourth Edition, Volume 5. Springer-Verlag, Berlin, 1976, pp. 703-706.
4. I. Krejcí and R. Zadina. *Planta Medica* 7: 1-7, 1959.
5. V. Schlemper, A. Ribas, M. Niccolau, and V. Cechinel Filho. *Phytomedicine* 3: 211-216, 1996.
6. *Bundesanzeiger,* February 1, 1990.

Horsetail

Art's perfect forms no moral need,
And beauty is its own excuse;
But for the dull and flowerless weed
Some healing virtue still must plead.

John Greenleaf Whittier
Songs of Labor
Dedication, Stanza 5

There are those who plead eloquently for the healing virtues of the "dull and flowerless weed" known as horsetail. *Equisetum arvense* L. (family Equisetaceae) is technically a pteridophyte and thus is more closely related to the ferns than to the flowering plants. Horsetail is a rushlike perennial with hollow, jointed stems and scalelike leaves, reproducing by means of spores, not seeds. The stems contain large amounts of silica and silicic acids (5 to 8 percent) that account for its use as a metal polisher and its synonym of souring rush. Several closely related species of *Equisetum* are similar to *E. arvense* in appearance and use. Another commonly used species is *E. hymale* L.

Enthusiasts call horsetail a valuable diuretic and astringent for treating various kidney and bladder ailments, ranging from kidney stones to cystic ulceration, and also recommend it as a rapid-acting remedy for dropsy. It is also called effective in treating tuberculosis, especially when accompanied by the "spitting of blood." External application is supposed to stop the bleeding of wounds and promote rapid healing.[1,2]

In addition to the silica compounds, horsetail contains about 5 percent of a saponin, designated equisetonin, and several flavone glycosides including isoquercitrin, galuteolin, and equisetrin.[3] A very small amount of nicotine (0.00004 percent) is also present.[4] The flavone glycosides and the saponin probably combine to account for the diuretic action of horsetail, which has been demonstrated experimentally but which is very slight.[5] There is no valid experimental evidence to support the hypothesis that the silica and silicic acid derivatives in the drug promote the healing of bleeding tubercular lesions in the lung.[6]

A caveat regarding safety is that several species of *Equisetum* have been implicated in livestock poisoning, particularly of horses. Known as equisetosis, a thiaminase (thiamine-destroying) activity has been confirmed in horses. Treatment involves massive doses of thiamine and, of course, removing the horsetail-containing hay from the diet. Health and Welfare Canada has required manufacturers to prove that *E. arvense* is free from thiaminase-like activity. A European species, *Equisetum palustre* L., is known to contain a toxic alkaloid, palustrine. The German Pharmacopoeia suggests that commercial supplies of *E. arvense* be examined for adulteration with other *Equisetum* species, notably, *E. palustre.*[7,8]

Even vigorous pleading does not produce much scientific support for the healing virtues of horsetail. The plant is a weak diuretic and little else.

REFERENCES

1. M. Grieve. *A Modern Herbal,* Volume 1. Dover Publications, New York, 1971, pp. 419-421.

2. H. Kreitmair. *Die Pharmazie* 8: 298-300, 1953.

3. H. A. Hoppe. *Drogenkunde,* Eighth Edition, Volume 2. Walter de Gruyter, Berlin, 1977, pp. 173-176.

4. J. D. Phillipson and C. Melville. *Journal of Pharmacy and Pharmacology* 12: 506-508, 1960.

5. H. Vollmer and K Hübner. *Nauyn-Schmiedebergs Archiv für experimentelle Pathologie und Pharmakologie* 186: 565-573, 592-605, 1937.

6. E. Steinegger and R. Hänsel. *Lehrbuch der Pharmakognoise,* Third Edition. Springer-Verlag, Berlin, 1972, p. 214.

7. A. Y. Leung and S. Foster. *Encyclopedia of Common Natural Ingredients Used in Food, Drugs, and Cosmetics,* Second Edition. John Wiley and Sons, New York, 1996, pp. 306-308.

8. N. W. Hamon and D. V. C. Awang. *Canadian Pharmaceutical Journal* 399-401: September, 1992.

Hydrangea

Hydrangea, or seven barks, consists of the underground portions (rhizome and roots) of the plant *Hydrangea arborescens* L. (family Saxifragaceae), an erect shrub growing in the eastern part of the United States from New York to Florida and west to Oklahoma.[1] It was originally used by the Cherokee Indians, who introduced it to the early settlers as a remedy for kidney stones. During the first few decades of this century, the drug saw some action for this and also as a diuretic in conventional medicine, but it lapsed into disuse until the recent revival of herbal remedies. Advocates of herbal medicine still recommend it for these conditions.[2]

There is no evidence, aside from empirical observations and anecdotes, that hydrangea has any therapeutic utility at all. But then, there have been no modern scientific studies of the drug's physiological activity and practically no investigations of its chemistry.[3] For example, a crystalline compound first isolated in 1887 and designated hydrangin remains chemically unidentified more than 100 years later. Much of the recent, albeit scant, scientific literature on hydrangea reflects reports of contact dermatitis from handling the leaves of various species in the genus.[4]

Hydrangea would probably not merit being included in this book—for there are certainly more effective diuretics and treatments for kidney stones—if it were not for the publicity surrounding another species of the same genus. The hydrangea ordinarily cultivated for its showy flowers is *Hydrangea paniculata* Siebold, in particular, a cultivar of that species designated Grandiflora. Its leaves have been smoked in a fashion analogous to marijuana to produce a kind of euphoria or "high." However, even the books devoted to such intoxicants emphasize that this practice "will either get one very stoned or very sick," as a result of a cyanide-producing compound contained in the leaves.[5]

There is no question that the practice can make the user very sick, but there is some doubt that the poisonous character of the leaves is due to cyanide. Reports dating back to the early 1900s did detect cyanide-producing compounds in the leaves of certain *Hydrangea* species, but the common ornamental variety was not among them. Hegnauer has concluded that probably only a few members of the genus may have this property and then in significant amounts only during the early stages of vegetative development.[6]

Still, the questionable identity of the toxic principle does not make smoking hydrangea leaves any more advisable or the illness that can result from it any less real. As a drug of use or abuse, the roots or leaves of any *Hydrangea* species have no merit except for the weak diuretic action of the underground parts of *H. arborescens.*

REFERENCES

1. H. W. Felter and J. U. Lloyd. *King's American Dispensatory,* Volume 2. The Ohio Valley Co., Cincinnati, Ohio, 1900, pp. 1000-1001.

2. N. Coon: *Using Plants for Healing,* Second Edition. Rodale Press, Emmaus, Pennsylvania, 1979, p. 122.

3. A. Y. Leung and S. Foster. *Encyclopedia of Common Natural Ingredients Used in Food, Drugs, and Cosmetics,* Second Edition. John Wiley and Sons, New York, 1996, pp. 309-310.

4. M. E. Kuligowski, A. Chang, and J. H. Leemreize. *Contact Dermatitis* 26: 269-270, 1992.

5. L. A. Young, L. G. Young, M. M. Klein, D. M. Klein, and D. Beyer. *Recreational Drugs.* Collier Books, New York, 1977 p. 99.

6. R. Hegnauer. *Chemotaxonomie der Pflanzen,* Volume 6. Birkhäuser Verlag, Basel, Switzerland, 1973, p. 326.

Hyssop

Almost all the write-ups on this plant in modern herbals include a few choice biblical quotations referring to its ritualistic use in cleansing people or places.[1,2,3] These quotations are extremely misleading, for it is very doubtful if the well-known garden herb *Hyssopus officinalis* L., which is now called "hyssop," has any relation whatever to the plant mentioned in the Old and New Testaments. Learned discussions regarding the identity of the biblical hyssop were underway some 250 years ago, and even then, more than eighteen different plants had been suggested.[4] The plant's true identity will probably never be established.

Even in the twentieth century, considerable confusion exists about the name hyssop. Therefore, we must affirm again that the plant discussed here under that title is just plain hyssop (*Hyssopus officinalis* L.) and not giant hyssop, hedge hyssop, prairie hyssop, or wild hyssop, all of which are entirely different species.

Hyssop is a perennial shrub of the family Lamiaceae that has been naturalized in the United States. It is a common garden plant that also grows widely along the sides of roads. As is the case with many other mints, its leaves and flowers contain an appreciable amount of volatile oil, giving them a camphorlike odor and a somewhat bitter taste. This volatile oil is an ingredient in many French liqueurs, specifically those which resemble Chartreuse and Benedictine.[5] It is also the agent responsible for the household medicinal use of the plant, mostly as tea, for coughs, colds, hoarseness, fevers, and sore throats.[6] Hyssop tea, mixed with a little honey, is said to be especially effective as an expectorant (an agent that promotes the loosening and expulsion of phlegm).

Because of the presence of pinocamphone, isopinocamphone, α- and β-pinene, camphene, and α-terpinene, which together make up about 70 percent of the oil,[7] hyssop is a reasonably effective treatment for mild irritations of the respiratory tract that accompany the common cold. It is also generally recognized as safe; no adverse reports of its safety appcar in thc scientific literature. Two recent studies have reported on preliminary anti-HIV activity from hyssop fractions. A California research group identified a polysaccharide (deemed MAR-10) that, depending upon concentration, inhibited the SF strain of HIV-1 in laboratory experiments designed to measure HIV-1 cell replication.[8,9]

On the other hand, claims of its ability to treat wounds or cuts (even those made with rusty farm implements), because the mold that produces penicillin grows on hyssop leaves, must be regarded as so much nonsense.[10] *Penicillium* species are among the most ubiquitous of fungi and grow anywhere there is sufficient moisture, nutrients, and a suitable temperature. It is highly unlikely that the sparse growths occurring on hyssop leaves would produce any viable amount of antibiotic. Any antiseptic action of the leaves would more likely be due to its volatile oil. Such an effect would be relatively weak at best and would certainly not be significant in the treatment of puncture wounds susceptible to tetanus (lockjaw) infection.

REFERENCES

1. R. Lucas. *Nature's Medicines.* Wilshire Book Co., North Hollywood, California, 1977, p. 25.

2. W. H. Hylton (Ed). *The Rodale Herb Book.* Rodale Press Book Div., Emmaus, Pennsylvania, 1976, pp. 474-478.

3. R. C. Wren and R. W. Wren. *Potter's New Cyclopaedia of Botanical Drugs and Herbs.* New Health Science Press, Hengiscote, England, 1975, p. 160.

4. H. N. Moldenke and A. L. Moldenke. *Plants of the Bible.* Chronica Botanica Company, Waltham, Massachusetts, 1952, pp. 160-162.

5. E. Guenther. *The Essential Oils,* Volume 3. D. Van Nostrand Company, New York, 1949, pp. 436-440.

6. W. H. Hylton (Ed.), op. cit., pp. 474-478.

7. A. Y. Leung and S. Foster. *Encyclopedia of Common Natural Ingredients Used in Food, Drugs, and Cosmetics*, Second Edition. John Wiley and Sons, New York, 1996, pp. 312-314.

8. W. Kreis, M. H. Kaplan, J. Freemand, D. K. Sun, and P. S. Sarin. *Antiviral Research* 14: 323-337, 1990.

9. S. Gollapudi, H. A. Sharma, S. Aggarwal, L. D. Byers, H. E. Ensley, and S. Gupta. *Biochemical and Biophysical Research Communications* 210: 145-151, 1995.

10. D. Hall. *The Book of Herbs.* Charles Scribner's Sons, New York, 1972, pp. 126-129.

Jojoba Oil

It is difficult to pick up a catalog of beauty aids these days and not find a section of it devoted to various preparations containing jojoba (pronounced ho-ho-ba) oil. Shampoos are most common, but various kinds of creams and lotions containing the "rare oil used by American Indians for hundreds of years as a cosmetic and medicinal aid" are also offered. Just what is this mysterious substance that seems to have become popular during the last two decades?

Jojoba wax (technically it is not an oil) is obtained from the peanut-sized seeds of the jojoba plant, *Simmondsia chinensis* (Link) C. K. Schneid., also referred to as *S. californica* (Link) Nutt. This evergreen shrub of the family Buxaceae grows in abundance on rocky desert hillsides in Arizona, California, and Mexico. Commonly called goat nuts, the seeds of the plant resemble coffee beans. When expressed, they yield about 50 percent of a liquid wax, composed almost entirely of high molecular weight, monoethylenic acids (eicosenoic acid—35 percent) and alcohols (eicosenol—22 percent, docosenol—21 percent). The wax, known commonly as jojoba oil, in its physical properties resembles sperm oil, formerly obtained from the sperm whale (now an endangered species).[1] It may be hydrogenated to produce a solid wax that is similar to spermaceti.

The American Indians reportedly used the oil as a hair dressing. Now it is incorporated into jojoba oil shampoos considered especially effective in preventing the buildup of sebum (the fatty material secreted by sebaceous glands) on the scalp. The reasoning behind this theory is that jojoba oil resembles sebum in many respects, both chemically and physically. By coating the scalp with jojoba oil, it is believed (but not proven) that the natural production of sebum will be reduced.[2] Of course, whether it is more desirable to have the scalp coated with jojoba oil or with sebum is a moot question. The oil is readily taken up by the skin and imparts a velvety softness to it and to the hair as well. This lubrication of the scalp does, no doubt, reduce the flaking of the skin normally associated with dandruff.

Overly enthusiastic advocates of jojoba oil, known in the trade as "jojoba witnesses," tout the product for restoring lost hair and preventing further hair loss, for removing warts, curing cancer, and other similar uses. Needless to say, there is absolutely no scientific evidence to support such claims.

Toxicity tests on jojoba oil for external application have caused no significant concern. Aside from causing occasional allergic responses in sensitive individuals, the product may be considered safe for human skin.[3] Whether the various preparations containing this unusual liquid wax are any more effective than similar cosmetics containing the customary emollient oils is something that will be very difficult to prove scientifically. As with most such products, this is best judged subjectively by each user.

REFERENCES

1. A. H. Warth. *The Chemistry and Technology of Wax.* Rheinhold Publishing Corp., New York, 1947, pp. 172-176.
2. T. K. Miwa. *Cosmetics and Perfumery* 88: 39-41, 1973.
3. J. H. Brown. *Manufacturing Chemist and Aerosol News* 50(6): 47, 1979.

Juniper

Persons of legal age who are not familiar with the odor and taste of juniper have led sheltered lives. For well over 300 years, gin has been one of the most popular alcoholic beverages of the Western world, most of whose inhabitants have some familiarity with the dry martini or other juniper-flavored concoctions. (So-called martinis made with tasteless vodka instead of gin are feeble substitutes for the real thing.) The fleshy, purplish "berries" (technically "cones" of *Juniperus communis* L. and its variety *depressa* Pursh, evergreen shrubs or small trees of the family Cupressaceae, contain 0.2 percent to 2 percent of a volatile oil that is the principal flavoring agent in gin.

Juniper and its volatile oil have long enjoyed a considerable reputation in folk medicine as a diuretic and as a treatment in various conditions of the kidneys and bladder. They are also recommended for their carminative action in cases of indigestion and flatulence.[1] The berries are usually taken in the form of a tea prepared from juniper in a mixture with other drugs, but for treating rheumatism, the berries themselves are eaten. The berries are also said to have a stimulating effect on the appetite; this may account for their incorporation as a flavoring agent in such dishes as sauerkraut. Over 105 constituents have been found in the essential oil, seventy-seven of which have been identified.[2] Extracting the berries with 70 percent alcohol yields a volatile oil-rich preparation, technically called a "spirit," which is suggested for either external or internal use in these conditions.[3]

The diuretic action of juniper results from its contained volatile oil and, specifically, the constituent designated terpinen-4-ol (1.37 percent of the oil by weight), which increases the glomerular filtration rate in the kidneys.[4] However, excessive doses of the drug may produce kidney irritation, and in the case of persons already suffering from kidney disease, this can result from even normal therapeutic doses. Juniper and its preparations must not be used by expectant mothers since they not only increase intestinal movements but also stimulate contraction of the uterus.[5] The concentration of juniper oil in commercial alcoholic beverages is quite small, not exceeding 0.006 percent,[6] so imbibers should not expect therapeutic responses when these are consumed (at least in reasonable amounts).

Because it acts as a diuretic by causing local irritation of the kidneys—and because this action is liable to be detrimental when those organs are already inflamed—and because juniper is hazardous for use by pregnant mothers, this drug is no longer recommended for various kidney disorders by the medical profession. Juniper berries could adversely influence glucose levels in diabetics. In European phytomedicine, use is limited to only one month, unless under a physician's supervision.[7] Safer and much more effective drugs certainly exist, but juniper continues to be used in folk medicine, particularly for its diuretic properties. In fact, a recent comprehensive review has challenged the assertion that juniper berries and their contained oil are nephrotoxic in therapeutic doses. Such effects may be due to products adulterated with turpentine oil.[8] Additional research on this ancient herbal remedy is certainly warranted.

REFERENCES

1. M. Grieve. *A Modern Herbal,* Volume 2. Dover Publications, New York, 1971, pp. 452-453.

2. P. S. Chatzoupoulou and S. T. Katsiotis. *Planta Medica* 59: 554-556, 1993.

3. M. Pahlow. *Das Grosse Buch der Heilpflanzen.* Gräfe und Unzer GmbH, Munich, Germany, 1979, pp. 341-342.

4. J. Janku, M. Hava, and O. Motl. *Experientia* 13: 255-256, 1957.

5. P. H. List and L. Hörhammer (Eds.). *Hagers Handbuch der Pharmazeutischen Praxis,* Fourth Edition, Volume 5. Springer-Verlag, Berlin, 1976, pp. 333-337.

6. A. Y. Leung and S. Foster. *Encyclopedia of Common Natural Ingredients Used in Food, Drugs, and Cosmetics*, Second Edition. John Wiley and Sons, New York, 1996, pp. 325-327.

7. ESCOP. Juniperi Frucutus. In *Monographs on the Medicinal Uses of Plant Drugs,* Volume 3. European Scientific Cooperative on Phytotherapy, Exeter, United Kingdom, 1997.

8. H. Schilcher and B. M. Heil. *Zeitschrift für Phytotherapie* 15: 205-213, 1994.

Kava

Kava refers to both the rootstock and traditional inebriating beverage derived from *Piper methysticum* Forest. f. (family Piperaceae). A sterile cultivated plant with numerous cultivars (cultivated varieties), it originates from a wild progenitor, *Piper wichmannii* C. DC., found from New Guinea to the Solomon Islands and Vanuatu. Kava is grown for its large rootstock or stump (often designated as a rhizome, and a matter of intense nomenclatural debate among botanists) cultivated on tropical Pacific Islands, including among others Baluan, Fiji, Futuna, Hawaii, Madang, Pohnpei, Rotuma, Samoa, Tonga, Vanuatu (formerly the New Hebrides), and Wallis. Vanuatu is considered the center of distribution since eighty of the known 118 cultivars of kava occur in this archipelago.[1]

Pacific Islanders normally consume kava beverages at dusk, before an evening meal. Various rituals or ceremonies are associated with the event, depending upon culture. The rootstock is prepared by chewing, grinding, grating, or pounding roots, then soaking and macerating the pulp in cold water to release the active constituents into what is often a thick brew, which has been offered to visiting dignitaries from Lady Bird Johnson to Pope John Paul II. In Pacific Island societies, kava consumption—the kava ceremony—has been likened to the social equivalent of consumption of wine in France.[2]

Kava's strong, somewhat nauseating taste produces localized numbing results from its complex chemistry, with kavapyrones (kavalactones) dominating flavor and biological activity. At least fifteen kavapyrones have been isolated and characterized, the most important of which are kawain (1 to 2 percent), dihydrokawain (0.6 to 1 percent), methysticin (1.2 to 2 percent), dihydromethysticin (0.5 to 0.8 percent),[3] in addition to demethoxyyangonin and yangonin, among others. The four kawain-methysticin-type pyrones act as muscle relaxants and anticonvulsants. Pipermethystine, an alkaloid, is a major constituent of the leaves, but is absent from the roots.[4,5]

Kava, as consumed in a typical Polynesian kava ceremony, is said to induce a state of mood elevation, well-being, and contentment, producing a feeling of relaxation without a narcotic effect. However, excessive consumption may cause photophobia and diplopia, sometimes resulting in oculomotor paralysis, with muscles not responding to normal movement, leading to prostration and unconsciousness. Heavy consumption over a period of several weeks or months can produce drying-up of the skin epidermis, with lesions and skin yellowing, redness of the eyes, loss of appetite, urticarial patches with intense itching, and other symptoms, which abate when kava consumption is discontinued.[6]

The first herb products made from kava appeared in Europe in the 1860s. At the end of the last century, kava extracts were available in German pharmacies. The first pharmaceutical preparations became available in Germany in the 1920s, offered as a mild sedative and hypotensive agent in the form of a tincture. For the most part, Western use of kava has largely evolved in Germany.

The rootstock and its preparations are currently the subject of a positive German therapeutic phytomedicine monograph, which allows it to be used for conditions of nervous anxiety, tension, and agitation. In Europe, kava extracts are often combined with pumpkin seed and used in the treatment of irritable bladder syndrome. The German monograph notes that use is contraindicated during pregnancy, lactation, and depression. Side effects may include temporary yellow discoloration of the skin, hair, and nails or rare allergic skin reactions. Because of its perceived sedative effect, use should be avoided with the consumption of alcohol or with operation of machinery or vehicles.[7,8] For these reasons, the herb is best consumed at bedtime. One case of driving under the influence of kava has been prosecuted in the state of Utah.

At least six double-blind controlled therapeutic studies have been conducted on kava extracts, though they have been criticized for insufficient inclusion criteria. Test substances have involved extracts standardized to 15 percent kavapyrones (two studies) and 70 percent kavapyrones (four studies). Clinical studies have measured various parameters in the treatment of anxiety, tension, agitation of nonpsychotic origin, climacteric symptoms, and postoperative mood with positive results. The data from controlled clinical studies, along with decades of clinical experience in Germany, can lead one to conclude that kava is a potential herbal alternative to benzodiazepines or other synthetic anxiolytics for mild states of anxiety from various causes. Although side effects of kava abuse are clearly established in the literature, it does not appear to produce physical or psychological dependency, an advantage over some synthetic anxiolytics and tranquilizers.[9]

Kava's tenure in the American market will depend upon responsible manufacturers offering appropriately formulated products in proper dosages that reduce the potential for abuse. They must also refrain from hyperbolic advertising featuring extravagant claims for the herb's psychotropic effects.

REFERENCES

1. V. Lebot, M. Merlin, and L. Lindstrom. *Kava: The Pacific Drug.* Yale University Press, New Haven, Connecticut, 1992, pp. 255.

2. Y. N. Singh and M. Blumenthal. *HerbalGram* 39: 33-55, 1997.

3. V. Schulz, R. Hänsel, and V. E. Tyler. *Rational Phytotherapy: A Physicians' Guide to Herbal Medicine,* Third Edition. Springer, Berlin, 1998, pp. 65-73.

4. Y. N. Singh and M. Blumenthal, op. cit., pp. 46-47.

5. V. Lebot. Kava (*Piper methysticum* Forst. f.): The Polynesian Dispersal of an Oceanian Plant. In *Islands, Plants and Polynesians: An Introduction to Polynesian Ethnobotany,* P. A. Cox and S. A. Banack (Eds.). Dioscorides Press, Portland, Oregon, 1991, pp. 169-201.

6. Ibid., p. 183.

7. *Bundesanzeiger,* June 1, 1990.

8. A. Y. Leung and S. Foster. *Encyclopedia of Common Natural Ingredients Used in Food, Drugs, and Cosmetics,* Second Edition. John Wiley and Sons, New York, 1996, pp. 330-331.

9. V. Schulz, R. Hänsel, and V. E. Tyler, op. cit., pp. 68-73.

Kelp

Kelp is such an imprecise generic term that no two authorities seem to agree on the identity of the plant(s) it supposedly designates. Definitions range all the way from a single species, *Fucus vesiculosus* L.[1] to any kind of coarse seaweeds, or more precisely, the ash obtained by burning them.[2] In modern times, kelp most probably refers to seaweeds of the brown algal order Laminariales, which possess large, flat, leaflike fronds, especially species of *Laminaria, Macrocystis,* and *Nereocystis.*[3] This definition should be broadened, however, to include several species of another order, Fucales, or rockweed, since several authors indicate that the brown algae known as bladderwrack, *Fucus vesiculosus,* is commonly used for producing kelp products.[4,5]

Kelp in the form of a powder or tablets is used in folk medicine to treat constipation, bronchitis, emphysema, asthma, indigestion, ulcers, colitis, gallstones, obesity, and disorders of the genitourinary and reproductive systems, both male and female. It is also claimed to "clean" the bloodstream, strengthen resistance to disease, overcome rheumatism and arthritis, act as a tranquilizer, combat stress, and alleviate skin diseases, burns, and insect bites.[6]

Many of these actions are attributed to kelp's content of minerals, especially iodine. But the concentration of iodine in seaweeds is extremely variable, not only among different species, but within the same species. For example, we find *Fucus vesiculosus* grown in the Baltic Sea contains 0.03 percent iodine; that harvested from the North Sea yields 0.1 percent, and still other specimens up to 0.2 percent.[7] Other species of *Fucus* may contain as much as 0.5 percent iodine. Consequently, if a kelp preparation is to be taken for its iodine content, it should be standardized, or at least it should be determined and stated on the label.

One of the advocated uses of kelp is to control obesity. This role is attributed to the plant's iodine content, which supposedly stimulates the production of iodine-containing thyroid hormones. However, such stimulation would only result in people suffering iodine deficiency, an almost unknown condition in this age of iodized salt. The required daily allowance of iodine in adults is quite small, not exceeding 150 micrograms; consequently, administering extra iodine beyond the ability of the thyroid to use it is essentially worthless in the short term. In the long term it may present some health risks. Even if it were effective, using increased amounts of thyroid hormones for weight reduction is not recommended.[8]

Claims are also made that iodine-containing kelp is useful as a blood vessel cleanser in the treatment of atherosclerosis. Iodine therapy for atherosclerosis is currently controversial and cannot be recommended. When it is employed, precise doses, which are not available in unstandardized kelp preparations, are administered.

Many of the other medicinal uses of kelp are dependent on its content of algin (sodium alginate), a high molecular weight polysaccharide found in all brown algae in concentrations ranging from about 12 to 45 percent.[9] Algin forms viscous, colloidal solutions or gels in water and is the ingredient in kelp responsible for its bulk laxative and demulcent (soothing) effects.

Actions that cannot be explained on the basis of kelp's iodine or algin content are probably not substantial. It is frequently pointed out by advocates that potassium is present in kelp in relatively large amounts. Although that is true, what is usually left unsaid is that the seaweed is also high in sodium (salt). Kelp should consequently be avoided by those who must restrict their salt intake.

One modern herbalist recommends the consumption of kelp to provide protection from the absorption of radioactive strontium 90.[10] Although this might have some merit in the event of a nuclear war or a major nuclear accident, such as that which occurred at Chernobyl, it certainly is not needed in this country today, nor can it be advocated as a routine preventive measure. The quantity of iodine in the suggested dose (three ounces per week) would far exceed the levels liable to induce or to intensify hyperthyroidism. For this reason, German health authorities do not recommend the use of kelp for any therapeutic purpose.[11] Another important caveat is that kelp may also concentrate heavy metals such as cadmium and lead from contaminated sea water. A recent epidemiological study among Canadian Inuit indigenous peoples found that kelp, along with other seafood normally consumed in the diet, contained relatively high levels of both cadmium and lead, prompting guidelines for monitoring heavy metal content of traditional seafoods, to help reduce health risks.[12] Besides, kelp tastes bad.

REFERENCES

1. *Stedman's Medical Dictionary,* Twenty-Third Edition. Williams and Wilkins, Baltimore, Maryland, 1976, p. 740.

2. *Encyclopedia Brittanica,* Volume 13. Encyclopedia Brittanica, Chicago, Illinois, 1952, p. 317.

3. V. J. Chapman. *Seaweeds and Their Uses,* Second Edition. Methuen and Co., Ltd., London, 1970.

4. G. J. Binding and A. Moyle. *About Kelp.* Thorsons Publishers Ltd., Wellingborough, England, 1974.

5. R. Lucas. *Common and Uncommon Uses of Herbs for Healthful Living.* Arco Publishing Company, New York, 1969, pp. 53-57.

6. G. J. Binding and A. Moyle, op. cit.

7. P. H. List and L Hörhammer (Eds.). *Hagers Handbuch der Pharmazeutischen Praxis,* Fourth Edition, Volume 4. Springer-Verlag, Berlin, 1973, pp. 1062-1065.

8. V. E. Tyler, L. R. Brady, and J. E. Robbers, *Pharmacognosy,* Ninth Edition. Lea and Febiger, Philadelphia, Pennsylvania, 1988, pp. 479-480.

9. V. J. Chapman, op. cit.

10. M. Castleman. *The Healing Herbs.* Rodale Press, Emmaus, Pennsylvania, 1991, pp. 230-231.

11. *Bundesanzeiger,* June 1, 1990.

12. H. M. Chan, C. Kim, K. Khoday, O. Receveur, and H. V. Kuhnlein. *Environmental Health Perspectives* 103: 740-746, 1995.

Lettuce Opium

This venerable fraud of a drug keeps coming back to confront us like the proverbial bad penny. Consisting of the dried milky juice or latex of several species of lettuce, it is collected from the stem of the plant, which is cut off at the time of flowering. The source most commonly utilized is the so-called wild lettuce, *Lactuca virosa* L., but garden lettuce, *L. sativa* L., as well as the related species, *L. serriola* L. and *L. sagittata* Waldst. & Kit., all members of the family Asteraceae, also yield the product. Lettuce opium is also known by the Latin title, lactucarium.[1]

Taken as a drug by the ancient Egyptians, lettuce opium was long thought to possess soporific (sleep-producing) properties; however, this was probably based on the similar appearance of the white milky juice exuded by the cut lettuce plant and that yielded by the opium poppy. The odor, taste, and general appearance of lactucarium also resemble those of opium.

Introduced into conventional American medicine in 1799 by J. R. Coxe, a Philadelphia physician, the use of lettuce opium as a sedative and painkiller flourished for a century or so and then gradually lost favor. By the mid-twentieth century, the drug had fallen into obscurity. Then suddenly, in the mid-1970s, it was resurrected as a legal psychotropic or mind-altering drug by members of the American hippie movement. Various lettuce opium preparations were widely advertised in counterculture publications, either as the pure material or combined with "potency enhancers" such as catnip and damiana. The products were intended to be smoked to produce a feeling of euphoria and well-being (a "high").[2] At the height of lettuce opium's popularity in 1977, one dealer was reported to be making $1,500 profit per *day* from the sale of extracted lettuce products.[3]

Over the years, repeated attempts have been made to demonstrate sedative and painkilling effects in lettuce opium and to identify active principles that might be responsible for them. As early as 1892, hyoscyamine was reported in extracts of various lettuce species but not in commercial lactucarium.[4] These observations have not been verified by subsequent investigations. An extensive pharmacological study of lettuce opium, published in 1940, showed that the fresh milky juice contained two bitter principles, lactucin and lactucopicrin, which had definite depressant or sedative effects on the central nervous system in small animals.[5] However, these compounds were found to be quite unstable, and commercial lactucarium had little, if any, activity.

An evaluation of lettuce opium in 1944 caused Fulton to reach the following conclusion: "Modern medicine considers its sleep producing qualities a superstition, its therapeutic action doubtful or nil."[6] More recently, Brown and Malone examined one of the modern lettuce opium preparations and concluded, "The analgesic, sedative and other attributes of lettuce opium lactucarium, seem to be based on fiction rather than fact."[7]

Although these conclusions are certainly factual, a rather startling revelation regarding the constituents of lettuce was made in 1981. Scientists reported that studies using an extremely sensitive radioimmunoassay technique detected minute amounts of morphine (2 to 10 nanograms per gram, dry weight) in both hay and lettuce.[8] However, before getting too excited over this discovery, remember that a nanogram is one billionth of a gram; also, similar small quantities of morphine were found in such unlikely natural sources as cow's milk and human milk. The amounts involved in either lettuce or milk would be far too small to exert any obvious physiological effect. Sensible people may continue to eat lettuce in their bacon and tomato sandwiches, but they will not smoke it in their pipes.

REFERENCES

1. M. Grieve. *A Modern Herbal,* Volume 2. Dover Publications, New York, 1971, pp. 476-477.

2. B. Rosen. *High Times* 23: 84, 1977.

3. R. K. Siegel. Street Drugs 1977: Changing Patterns of Recreational Use. In *Drug Abuse and Alcoholism,* S. Cohen (Ed.). The Haworth Press, Inc., Binghamton, New York, 1981, p. 12.

4. T. S. Dymond. *Journal of the Chemical Society,* Transactions 61: 90-94, 1892.

5. A. W. Forst. *Naunyn-Schmiedebergs Archiv für experimentelle Pathologie und Pharmakologie* 195: 1-25, 1940.

6. C. C. Fulton. *The Opium Poppy and Other Poppies.* Bureau of Narcotics, U.S. Treasury Dept., U.S. Government Printing Office, Washington, DC, 1944, pp. 62-63.

7. J. K. Brown and M. H. Malone. *Pacific Information Service on Street Drugs* 5(3-6): 36-38, 1977.

8. E. Hazum, J. J. Sabatka, K.-J. Chang, D. A. Brent, J. W. A. Findlay, and P. Cuatrecasas. *Science* 213: 1010-1012, 1981.

Licorice

Can we ever have too much of a good thing?

Signor Licentiate Pero Perez
in *Don Quixote*

Licorice consists of the underground parts, technically the rhizome and roots, of varieties of the European species *Glycyrrhiza glabra* L. (family Fabaceae), which possess a sweet yellow wood. Since significant quantities of licorice in the American market is imported from China, *Glycyrrhiza uralensis* Fisch. and other Asian species are also involved in the commercial supply. The drug is often called licorice root or glycyrrhiza. The root has been used since very ancient times as a flavoring and for its expectorant and demulcent properties in the treatment of coughs and colds.[1]

A very confusing situation exists concerning much so-called licorice candy. A great deal of it contains little or no licorice whatsoever but derives its flavor from anise oil. The flavors of licorice and anise do resemble each other, but in other respects, including potential toxicity, they are quite different. It is unfortunate that the more common, harmless anise flavor is almost always referred to as licorice.

Millions of pounds of licorice are imported into the United States annually, most of it originating in the eastern Mediterranean region. About 90 percent of it is used in flavoring tobacco products—cigarettes, cigars, pipe tobaccos, and the like. The amounts used are very closely guarded trade secrets, but the noticeable sweetness and pleasant flavor of many commercial tobacco blends is due to licorice. It is also an ingredient in various pharmaceuticals, especially throat lozenges. Authentic licorice candy is far more popular in Europe, especially Britain, than in this country.

Much of the sweetness of licorice is due to glycyrrhizin, also known as glycyrrhizic acid, a saponin glycoside that occurs in the root in concentrations averaging between 5 and 9 percent. It is about fifty times sweeter than sugar and is available commercially in a form known as ammoniated glycyrrhizin.

During World War II, a Dutch physician noted that administration of licorice extract produced marked improvement in patients suffering from peptic ulcer, but serious side effects, in the form of swelling of the face and limbs, were also observed.[2] Since then, numerous reports of toxic effects have been recorded in the medical literature, based on the observation of patients who ate large amounts of licorice candy over long periods of time. One man who had eaten two or three 36 gram licorice candy bars daily for six to seven years became so weak he could not get out of bed. He required hospitalization with intensive treatment for more than one month before recovering.[3] Another person, previously in excellent health, ate 700 grams (about 1.5 pounds) of licorice candy in a nine-day period. His condition necessitated four days of hospital treatment.[4] In a controlled experiment, about 100 to 200 grams of licorice twists (equivalent to 0.7 to 1.4 grams of glycyrrhizin), eaten daily for periods of one to four weeks, produced serious symptoms in a group of volunteers.[5]

Another interesting case involved an elderly man who chewed eight to twelve three-ounce bags of chewing tobacco daily and swallowed the saliva produced. He became so weak that he was unable to sit up or raise his arms above the horizontal position. There was prompt improvement when he was hospitalized and denied chewing tobacco. Tests revealed that the brand he had been consuming contained more than 8 percent licorice paste and that his usage amounted to between 0.88 and 1.33 grams of glycyrrhizin per day, well within the toxic range.[6]

The medical literature refers to this condition as pseudoaldosteronism, meaning one similar to that brought about by excessive secretion of the adrenal cortex hormone, aldosterone. In the case of licorice, the syndrome is caused by glycyrrhizin, the structure and physiological effects of which are related to aldosterone or desoxycorticosterone. Symptoms resulting from excessive quantities include headache, lethargy, sodium and water retention, excessive excretion of potassium, high blood pressure, and even heart failure or cardiac arrest.[7]

One popular herbal cough remedy contains 1 ounce of licorice root in a quart of water. Directions suggest that one-half pint be drunk at bedtime with additional quantities as needed.[8] The half-pint dose could easily contain 0.5 gram of glycyrrhizin, and that daily amount might be doubled or tripled, depending on the frequency of use. At that rate of consumption, toxic effects could be observed after a single week. For persons suffering from high blood pressure or heart trouble, these could be serious.

Since well-documented toxicity is associated with glycyrrhizin, deglycyrrhizinated licorice products have appeared in the American market in recent years. These may have some residual activity, but of course, the principal active constituent has been removed.

Although licorice does have a flavor pleasing to many and may also have some utility in treating coughs as well as a number of other conditions,[9] it must be remembered that it is also a potent botanical. Large doses over extended periods of time are quite toxic.[10]

When it comes to licorice, the answer to Pero Perez's question is "Yes!"

REFERENCES

1. V. E. Tyler, L. R. Brady, and J. E. Robbers. *Pharmacognosy,* Ninth Edition. Lea and Febiger, Philadelphia, Pennsylvania, 1988, pp. 68-69.

2. C. Nieman. *Chemist and Druggist* 177: 741-745, 1962.

3. J. W. Conn, D. R. Rovner, and E. L. Cohen. *Journal of the American Medical Association* 205: 492-496, 1968.

4. T. J. Chamberlain. *Journal of the American Medical Association* 213: 1343, 1970.

5. M. T. Epstein, E. A. Espiner, R. A. Donald, and H. Hughes. *British Medical Journal* 1: 488-490, 1977.

6. J. D. Blachley and J. P. Knochel. *New England Journal of Medicine* 302: 784-785, 1980.

7. Anon. *Medical Letter on Drugs and Therapeutics* 21(7): 30, 1979.

8. R. C. Wren and R. W. Wren. *Potter's New Cyclopaedia of Botanical Drugs and Preparations,* New Edition. Health Science Press, Hengiscote, England, 1975, p. 187.

9. R. F. Chandler. *Canadian Pharmaceutical Journal* 118: 420-424, 1985.

10. *Lawrence Review of Natural Products,* June, 1989.

Life Root

All of the old-time beliefs about the medicinal properties of the plant *Senecio aureus* L. are so aptly summarized in two sentences contained in a turn-of-the-century "doctor" book that it is worthwhile to quote them here:

> Life-root exerts a peculiar influence upon the female reproductive organs, and for this reason has received the name of Female Regulator. It is very efficacious in promoting the menstrual flow, and is a valuable agent in the treatment of uterine diseases.[1]

A modern herbal, which provides much the same information, includes the statement, "A most valuable herb indeed."[2]

Life root, also known as golden senecio, ragwort, false valerian, and squaw weed is a perennial herb with bright yellow flower heads belonging to the family Asteraceae. It grows in swampy grounds and moist thickets throughout the eastern and central United States. Actually, the entire dried plant, not just the root, was used as a drug.[3] In 1979, it was still one of the principal ingredients in that famous old proprietary remedy Lydia Pinkham's Vegetable Compound.[4]

Small but readily detectable quantities (0.006 percent) of the toxic alkaloid senecionine are now known to be present in life root.[5] Senecionine belongs to the group of hepatotoxic (poisonous to the liver) pyrrolizidine alkaloids that are effective in inducing chronic disease in rats with one, or at most, a few doses.[6] A strong possibility exists that such alkaloids are involved in human liver diseases, including primary liver cancer.

For this reason, it is not useful to discuss in detail the therapeutic potential of life root or the desirability of self-medicating with it for conditions in which it may be effective. Because of the presence of senecionine, the herb is simply not safe to use. Its employment in herbal medicine should be discontinued as was its use, many years ago, in more conventional therapy.

REFERENCES

1. R. V. Pierce. *The People's Common Sense Medical Adviser in Plain English: or, Medicine Simplified,* Sixty-Second Edition. World's Dispensary Printing Office and Binding, Buffalo, New York, 1895, pp 341-342.

2. F. and V. Mitton. *Mitton's Practical Modern Herbal.* W. Foulsham and Co., London, 1976, p. 103.

3. H. W. Youngken. *Textbook of Pharmacognosy,* Sixth Edition. The Blakiston Co., Philadelphia, Pennsylvania, 1948, pp. 889-891.

4. S. Stage. *Female Complaints.* W. W. Norton and Co., New York, 1979, p. 89.

5. R. H. F. Manske and H. L. Holmes. *The Alkaloids,* Volume 1. Academic Press, New York, 1950, p. 159.

6. R. Schoental. In *Chemical Carcinogens,* C. E. Searle (Ed.). American Chemical Society, Washington, DC, 1976, pp. 626-689.

Linden Flowers

A number of different species of the genus *Tilia* yield flowers that are used in folk medicine, but most of the commercial product is derived from *T. cordata* Mill. and *T. platyphyllos* Scop. Also known as basswood or lime trees, these are large deciduous trees of the family Tiliaceae, which often grow over 100 feet in height. The former species is commonly referred to as the small-leaved European linden, the latter as the large-leaved linden. After collection in late spring, the yellowish or white fragrant flowers are rapidly dried in the shade. They must also be carefully preserved since even a small amount of moisture reduces their aromatic properties and their activity.[1]

Linden flower tea has been used since the late Middle Ages as a diaphoretic, that is, a drug which promotes perspiration. It is also recommended as both a nervine (tranquilizer) and a stimulant, two quite contradictory uses. In addition, it is considered valuable in the treatment of headaches, indigestion, hysteria, and diarrhea. Linden flowers were once thought to be so effective in the treatment of epilepsy that a patient could be cured simply be sitting under the tree![2]

A number of flavonoid compounds, particularly derivatives of quercetin and kaempferol, are found in linden flowers and, together with *p*-coumaric acid, are apparently responsible for the drug's diaphoretic properties. A pleasant-smelling volatile oil also occurs in the flowers, along with quantities of tannin and mucilage.

Studies have shown that the relative amounts of tannin and mucilage are extremely important as far as the taste of linden flower tea is concerned. The taste becomes significant because relatively large amounts need to be drunk to induce perspiration. Flowers with a high tannin (2 percent or greater) and relatively low mucilage content produce a more tasty tea than those with a lower concentration of tannin and large amounts of mucilage. The latter tend to be quite insipid. This explains why the flowers of *T. cordata* and *T. platyphyllos* are preferred sources of the herb. They contain relatively more tannin and less mucilage than the flowers of such species as *T. tomentosa* Moench, the silver linden. Consequently, teas prepared from the first two species taste much better.[3]

Authorities generally agree that linden flower tea is both a pleasant-tasting beverage and a useful diaphoretic.[4] The claims of therapeutic efficacy for other conditions should be disregarded. For the best-tasting product, select the flowers obtained from either *T. cordata* or T. *platyphyllos.* This may not be too easy if one has to rely on commercial sources, which often fail to identify the botanical source. The flowers should also be stored in airtight, light-resistant containers to preserve their maximum fragrance.

There is no evidence to support the statement by Grieve that tea from very old linden flowers may produce symptoms of narcotic intoxication.[5] However, it has been reported that too-frequent use of linden flower tea may result in damage to the heart.[6] Although this apparently occurs only occasionally and as a result of excessive intake of the beverage, those with known cardiac problems would do well to avoid using the herb.

REFERENCES

1. P. H. List and L Hörhammer (Eds.). *Hagers Handbuch der Pharmazeutischen Praxis,* Fourth Edition, Volume 6C. Springer-Verlag, Berlin, 1979, pp. 180-184.
2. R. C. Wren and R. W. Wren. *Potter's New Cyclopaedia of Botanical Drugs and Preparations,* New Edition. Health Science Press, Hengiscote, England, 1975, p. 184.
3. M. Luckner, O. Bessler, and P. Schröder. *Pharmazeutische Zentralhalle* 104: 641-644, 1965.
4. E. Steinegger and R. Hänsel. *Lehrbuch der Pharmakognosie und Phytopharmazie,* Fourth Edition. Springer-Verlag, Berlin, 1988, pp. 123-124.
5. M. Grieve. *A Modern Herbal,* Volume 2. Dover Publications, New York, 1971, pp. 485-486.
6. M. Pahlow. *Das Grosse Buch der Heilpflanzen.* Gräfe und Unzer GmbH, Munich, Germany, 1979, pp. 221-223.

Lobelia

Throughout the nineteenth century, lobelia was such a favorite remedy of the practitioners of eclectic medicine that they were often called "lobelia doctors," a derogatory label. *Lobelia inflata* L. (family Campanulaceae), with its erect branched stem, alternate ovate or oblong leaves, and small, pale blue flowers, is quite an attractive plant. It is relatively common in the open woods and meadows of eastern North America but is also widely cultivated. The dried leaves and tops of the plant, also called Indian tobacco, make up the herb.[1]

Lobelia was valued primarily as a nauseant expectorant in cases of asthma and chronic bronchitis. In large doses, it acts as an emetic. Herbalists also report a beneficial action in conditions ranging from tuberculosis and nervous disorders to drug withdrawal,[2] but the drug's effectiveness in such cases is more imaginary than real. The renewed popularity of lobelia in the 1970s, and its consequent inclusion in all the modern writings on herbs, stems from its reputation as a euphoriant. Members of the counterculture smoke lobelia to obtain a mild, legal "high" analogous to that produced by smoking marijuana. The same feelings of mental clarity, happiness, and well-being are supposedly obtained by drinking lobelia tea or taking capsules of the powdered drug.[3]

The physiological activity of lobelia is accounted for by a mixture of pyridine-derived alkaloids contained in the plant to the extent of about 0.48 percent. Lobeline, lobelanine, and lobelanidine are the principal ones, with lobeline accounting for most of the drug's effects.[4] Less potent than nicotine but otherwise similar in its physiological action, lobeline first excites the central nervous system and then depresses it. In normal doses, it produces dilation of the bronchioles and increased respiration, but overdoses result in respiratory depression, as well as a host of other undesirable side effects, including sweating, rapid heartbeat, low blood pressure, and even coma followed by death.

Large doses may cause convulsions.[5] A recent preliminary pharmacological study in mice found antidepressant activity in a crude methanolic extract of the leaves, with beta-amyrin palmitate identified as an active component.[6] Lobelia's value as an expectorant and treatment for asthma and bronchitis is limited by the tendency of the drug to upset the stomach. It is an effective emetic when administered in adequate quantity, but it is really not safe enough to use.[7]

Because of the similarity of its physiological effects to those of nicotine, lobeline was formerly incorporated into tablets, lozenges, or pieces of chewing gum usually containing 2 to 4 mg of lobeline sulfate, which are claimed to help people stop smoking by masking the withdrawal symptoms of nicotine addiction. Unfortunately, the results of controlled trials with the drug have been disappointing.[8] Several of these products are nevertheless still marketed.

Self-administration in any form (smoking, drinking, eating) of inexact amounts of any crude drug as potent as lobelia is inadvisable. The ratio of risk to benefits is very high, especially when one considers the availability of much safer and effective treatments for any condition for which lobelia might conceivably prove helpful. Even the value of known quantities of lobeline as a smoking deterrent has not been proven. In short, the crude drug has been thoroughly discredited, and any use of it, either for therapeutic purposes or as a so-called euphoriant, is definitely not recommended.

REFERENCES

1. H. W. Felter and J. U. Lloyd. *King's American Dispensatory,* Eighteenth Edition, Volume 2. The Ohio Valley Co., Cincinnati, Ohio, 1900, pp. 1199-1205.

2. L. Griffin. *Herbalist* 2(7): 4-7, 1977.

3. L. A. Young, L. G. Young, M. M. Klein, D. M. Klein, and D. Beyer. *Recreational Drugs.* Collier Books, New York, 1977, p. 114.

4. A. Y. Leung and S. Foster. *Encyclopedia of Common Natural Ingredients Used in Food, Drugs, and Cosmetics,* Second Edition. John Wiley and Sons, New York, 1996, pp. 354-355.

5. A. Wade (Ed.). *Martindale: The Extra Pharmacopoeia,* Twenty-Seventh Edition. The Pharmaceutical Press, London, 1977, p. 313.

6. A. Subarnas, Y. Oshima, Y. Sidik, and Y. Ohizumi. *Journal of Pharmaceutical Sciences* 81: 620-621, 1992.

7. T. Sollmann. *A Manual of Pharmacology,* Seventh Edition. W. B. Saunders, Philadelphia, Pennsylvania, 1948, p. 352.

8. A. Y. Leung and S. Foster, op. cit., pp. 354-355.

Lovage

A tall perennial herb with dark green leaves and greenish yellow flowers, *Levisticum officinale* W. D. J. Koch (family Apiaceae) is extensively cultivated throughout much of Europe and the United States. In Europe, it is commonly known as the Maggi plant, sharing that name with the very popular, piquant flavoring sauce in which it is an important ingredient. All parts of the plant are highly aromatic, with an odor and taste reminiscent of celery. The leaves are often used as a seasoning, especially for soups; the underground parts (rhizome and roots) constitute the drug.[1]

Since earliest times, particularly since the fourteenth century, lovage root has been a celebrated folk medicine, primarily for its diuretic and carminative properties. Besides increasing the flow of urine and expelling gas, it was thought to be of value for treating kidney stones, jaundice, malaria, sore throat, pleurisy, and boils.[2] Pahlow tells us that the root is still used in Europe for the self-treatment of stomach upsets, bladder and kidney problems, rheumatism, gout, menstrual disturbances, and migraine headaches.[3] Misled by the name lovage, many persons value the plant as an ingredient in love potions.

Lovage root contains 0.6 to 1.0 percent of a volatile oil, the principal constituents (70 percent) of which are a series of lactone derivatives known as phthalides. Some of the same constituents are also present in celery seed, thus explaining the similarities in taste and odor. Injection into rabbits and mice of small quantities of lovage extract and also of lovage volatile oil resulted in pronounced diuresis. Some dermatitis due to photosensitivity was also observed.[4] Sensitivity to light is a common phenomenon of apiaceous plants, as we have already mentioned with angelica. It apparently is not a serious problem, for lovage continues to be sold in Europe as the principal ingredient (some say the only active ingredient) in many diuretic tea mixtures. Lovage is an important flavoring ingredient, too, in various liqueurs, herb bitters, and sauces.

The bulk of the evidence would indicate that lovage root does possess a definite diuretic action. It may also be helpful in relieving flatulence or gas. Agents that do this (carminatives), according to one of my past pharmacy professors, are drugs which "give the patient a good burp." There is no question that fresh lovage leaves or roots, chopped fine and cooked in various meat soups, stews, or the like, definitely enhance the flavor.

REFERENCES

1. E. Steinegger and R Hänsel. *Lehrbuch der Pharmacognosie,* Third Edition. Springer-Verlag, Berlin, 1972, p. 427.

2. M. Grieve. *A Modern Herbal,* Volume 2. Dover Publications, New York, 1971, pp. 499-500.

3. M. Pahlow. *Das Grosse Buch der Heilpflanzen.* Gräfe und Unzer GmbH, Munich, Germany, 1979, pp. 219-221.

4. P. H. List and L. Hörhammer (Eds.). *Hagers Handbuch der Pharmazeutischen Praxis,* Fourth Edition, Volume 5. Springer-Verlag, Berlin, 1976, pp. 497-500.

Milk Thistle

Milk thistle, also known as the Marian, St. Mary's, or Our Lady's thistle, is a tall herb with prickly leaves and a milky sap. It is native to the Mediterranean region of Europe but naturalized in California and the eastern United States. Botanically, the plant is known as *Silybum marianum* (L.) Gaertn., a member of the family Asteraceae. In older literature, as well as some modern European works, it is cited as *Carduus marianus* L. Over the years, several other plants have been referred to as milk thistles, but authorities now reserve that common name for this species. Also, it must not be confused with the blessed or holy thistle, which is *Cnicus benedictus* L., an entirely different plant, although the similarity of the religiously inspired common names is confusing.[1]

Another area of confusion with respect to the milk thistle is the part used; these are small hard fruits known technically as achenes from which a feathery tuft or pappus has been removed. Most of the English language herbal literature incorrectly refers to these fruits as seeds, which they do resemble, but which they are not. To confuse the matter more, some products have appeared in the market in recent years that contain milk thistle leaf. No therapeutic efficacy can be expected from milk thistle leaf–containing products. Virtually all research has been conducted on the fruits, specifically a well-defined extract of the fruits.

The fruits of the milk thistle have been used for many years for a variety of conditions, but especially for liver complaints. However, medicinal use of the plant, except perhaps as a simple bitter, was practically discontinued early in the twentieth century. In 1947, the *United States Dispensatory* devoted one short paragraph to the drug, primarily to its historical aspects.[2]

Then, about thirty years ago, German scientists undertook a chemical investigation of the fruits and succeeded in isolating a crude mixture of antihepatotoxic (liver protectant) principles designated silymarin, which is contained in the fruits in concentrations ranging from

1 to 4 percent. Subsequently, silymarin was shown to consist of a large number of flavonolignans, including principally silybin, accompanied by isosilybin, dehydrosilybin, silydianin, sily-christin, and others.[3]

Studies in small animals have shown that silymarin exerts a liver protective effect against a variety of toxins, including the phallotoxins of the deadly amanita, and is considered the only antidote to amanita poisoning.[4] Human trials have also been encouraging for conditions including hepatitis and cirrhosis of various origins.[5] The results of numerous studies suggest that silymarin has considerable therapeutic potential, protecting intact liver cells, or cells not yet irreversibly damaged, by acting on the cell membranes to prevent the entry of toxic substances. Protein synthesis is also stimulated, thereby accelerating the regeneration process and the production of liver cells. As a result of this information, German health authorities have endorsed the use of the herb as a supportive treatment for inflammatory liver conditions and cirrhosis.[6]

Unfortunately, silymarin is very poorly soluble in water, so the herb is not effective in the form of a tea. Studies show that such a beverage contains less than 10 percent of the initial activity in the plant material.[7] This poor solubility, coupled with the fact that silymarin is relatively poorly absorbed (20 to 50 percent) from the gastrointestinal tract, make it obvious that the active principles are best administered parenterally, that is, by injection. Oral use requires a concentrated product. Milk thistle is marketed in this country as a dietary supplement in the form of capsules containing 200 mg of a concentrated extract representing 140 mg of silymarin. Toxic effects resulting from the consumption of milk thistle have apparently not been reported. Twenty-one cases out of 2,169 (1 percent) in an observational study did report transient gastrointestinal side effects. Otherwise, it is considered very well-tolerated and quite effective.[8]

REFERENCES

1. R. F. Weiss. *Herbal Medicine.* Beaconsfield Publishers Ltd, Beaconsfield, England, 1988, pp. 82-85.

2. A. Osol and G. E. Farrar Jr. (Eds.). *The Dispensatory of the United States of America*, Twenty-Fourth Edition. J. B. Lippincott Company, Philadelphia, Pennsylvania, 1947, p. 1629.

3. H. Wagner and O. Seligmann. Liver Therapeutic Drugs from *Silybum marianum.* In *Advances in Chinese Medicinal Materials Research,* H. M. Chang, H. W. Yeung, W. W. Tso, and A. Koo (Eds.). World Scientific Publishing Co., Singapore, 1985, pp. 247-256.

4. S. Foster. *Milk Thistle,* Silybum marianum, *Botanical series No. 305,* Second Edition. American Botanical Council, Austin, Texas, 1996, pp. 7.

5. P. Ferenci, B. Dragosics, H. Dittrich, H. Frank, L. Benda, H. Lochs, S. Meryn, W. Base, and B. Schneider. *Journal of Hepatology* 9: 105-113, 1989.

6. *Bundesanzeiger,* March 13, 1986.

7. I. Merfort and G. Willuhn. *Deutsche Apotheker Zeitung* 125: 695-696, 1985.

8. V. Schulz, R. Hänsel, and V. E. Tyler. *Rational Phytotherapy: A Physicians' Guide to Herbal Medicine,* Third Edition. Springer, Berlin, 1998, pp. 214-220.

Mistletoe

The word "mistletoe" is about as nonspecific a term as you could possibly apply to a plant material. The addition of "American" or "European" helps a little. When properly used, American mistletoe refers to a single one of the more than 200 species of the genus *Phoradendron.* However, this species has four different scientific names, each of which is used more or less interchangeably. The most acceptable designation, of quite recent coinage, is *Phoradendron leucarpum* (Raf.) Rev. & M. C. Johnst., synonymous with *P. serotinum* (Raf.) M. C. Johnst. and *P. flavescens* (Pursh) Nutt. The nomenclature seems to be evolving more rapidly than the plant group itself. Once considered synonymous with *P. tomentosum* (DC.) Engelm, subspecies *macrophyllum* (Cockerell) Wiens, that taxon is now referred to as *P. macrophyllum* (Engelm.) Cockerell subspecies *macrophyllum.*[1,2] At first glance, the nomenclature for European mistletoe seems simpler; it is *Viscum album* L. But there are three subspecies commonly recognized: *platyspermum* Kell., growing on broadleaf trees; *abietis* Beck, growing on silver fir; and *laxum* Fiek, growing on various pines, seldom on firs.[3] All of these plants are parasitic shrubs belonging to the family Viscaceae.

Although the berries of both American and European mistletoe have long been considered poisonous, the leaves, in the form of a tea, have a considerable reputation as a home remedy. The reputed uses of the two plants are as different as their names. American mistletoe is believed to stimulate smooth muscles, causing a rise in blood pressure and increased uterine and intestinal contractions. European mistletoe has precisely the opposite reputation of reducing blood pressure and acting as an antispasmodic and calmative agent.[4]

Actually, both kinds of mistletoe contain toxic proteins that are very similar in their chemical composition.[5] These are designated phoratoxin when isolated from *Phoradendron* species and viscotoxins when obtained from various subspecies of *Viscum album.* Contrary to the folkloric reputation of the respective plants containing them, phoratoxin and the viscotoxins produced similar effects when injected into test animals. These included hypotension, slowing and weakening of the heartbeat, and constriction of the blood vessels in the skin and skeletal muscles.[6] However, it must be noted that the effects of these toxins following oral administration in human beings have not been studied.

Extracts of European mistletoe are sometimes employed in Germany in the treatment of malignant tumors. The Austrian spiritualist, Rudolf Steiner, introduced the European mistletoe to alternative cancer therapy in the early part of this century, based on "intuition" rather than any clinical evidence.[7] A sterile solution, available commercially, is injected either intravenously or into the tumor itself to provide palliative treatment for certain types of cancer.[8] The drug has not been approved for use in the United States. Such use of mistletoe extracts has led to identification in the plant of three lectins, that is, proteins which agglutinate red blood cells.[9] Many plant lectins are highly cytotoxic, and research is currently being conducted to determine their potential in cancer chemotherapy.

Certain Australian species of mistletoe have been shown to extract toxic principles, such as alkaloids and glycosides, from the host plants on which they grow as parasites. Thus, mistletoes grown on *Duboisia* species contain toxic solanaceous alkaloids and those grown on oleander contain potent cardiac glycosides.[10] The identity of the host plants on which the parasitic mistletoe is found is therefore extremely important if the crude plant material is to be used as a medicine.

Many popular writers on herbs recommend mistletoe tea as a treatment for conditions from anxiety to cancer. Because of the relatively high price of coffee, some persons have even advocated it as a pleasant-tasting substitute. Recent surveys of poisonous plants in the United States continue to emphasize the toxic nature of American mistletoe berries,[11] but German sources now maintain that the berries, but not the leaves, of European mistletoe have only slight toxicity, if any.[12,13] Until more definitive information is forthcoming, use of either type of mistletoe as a home remedy or as a beverage should definitely be avoided.

REFERENCES

1. J. L. Reveal and M. C. Johnston. *Taxon* 38: 107-108, 1989.

2. J. T. Kartesz. *A Synonymized Checklist of the Vascular Flora of the United States, Canada, and Greenland,* Second Edition, Volume 2. Timber Press, Portland, Oregon, pp. 585-586.

3. H. Becker and H. Schmoll. *Mistel.* Wissenschaftliche Verlagsgesellschaft mbH, Stuttgart, Germany, 1986, 132 pp.

4. P. S. Brown. *Herbalist* 1(12): 449-450, 1976.

5. S. T. Mellstrand and G. Samuelsson. *European Journal of Biochemistry* 32: 143-147, 1973.

6. S. Rosell and G. Samuelsson. *Toxicon* 4: 107, 1966.

7. H.-J. Gabius and S. Gabius. Phytotherapeutic Immunomodulation As a Treatment Modality in Oncology: Lessons from Research with Mistletoe. In *Phytomedicines of Europe Chemistry and Biological Activity.* L. D. Lawson and R. Bauer (Eds.). American Chemical Society, Washington, DC, 1998, pp. 278-286.

8. H. Becker and G. Schwarz. *Deutsche Apotheker Zeitung* 112: 1462-1465, 1972.

9. R. A. Locock. *Canadian Pharmaceutical Journal* 119: 124-127, 1986.

10. C. Boonsong and S. E. Wright. *Australian Journal of Chemistry* 14: 449-457, 1961.

11. W. H. Blackwell. *Poisonous and Medical Plants.* Prentice-Hall, Englewood Cliffs, New Jersey, 1990, pp. 241-242.

12. D. Frohne and H. J. Pfänder. *A Colour Atlas of Poisonous Plants.* Wolfe Publishing, Ltd., London, 1984, pp. 155-156.

13. J. L. Reveal and M. C. Johnston, op. cit., pp. 107-108.

Mormon Tea

Mormon tea is prepared from the fresh or dried stems of *Ephedra nevadensis* Wats., family Ephedraceae, a small erect shrub native to the desert regions of the southwestern United States and adjacent parts of Mexico. Twelve species of *Ephedra* occur in the southwestern deserts of the United States.[1] It is likely that many of them are harvested, without regard to species identity, as "Mormon tea." Many have not been studied chemically or pharmacologically. Called popotillo by the Mexicans, and Mormon tea, Brigham tea, teamster's tea, or squaw tea by the early American settlers, it was once a very popular folk remedy for syphilis and, especially, gonorrhea. Although its taste is quite astringent, those who become accustomed to it enjoy it as a pleasantly refreshing beverage.[2] The name Mormon tea probably derives from its use as a caffeine-free thirst quencher.

Since 1552, the plant yielding Mormon tea has been recommended as being beneficial to health. Widely used by frontiersmen as a cure for venereal disease, it is also described as a remedy for colds and kidney disorders and as a "spring tonic."[3] Spoerke attributes its activity to the presence of an undetermined amount of the alkaloid ephedrine, a drug that constricts the blood vessels, dilates the bronchioles, and stimulates the central nervous system.[4] Gottlieb,[5] Mowrey,[6] and Castleman[7] state that Mormon tea's active constituent is not ephedrine but (+)-norpseudoephedrine, an even more potent central nervous system stimulant.

Actually, five different groups of investigators have been unable to detect the presence of ephedrine, (+)-norpseudoephedrine, or any other alkaloid in *E. nevadensis,* and we may safely conclude that the plant is alkaloid free.[8] This is in keeping with all other North American species of *Ephedra,* which are singularly devoid of alkaloids. Mormon tea does contain large amounts of tannin, in addition to a resin and a volatile oil.

Administration of a fluidextract and an infusion (tea) of the drug to human subjects produced definite, but relatively mild, diuresis. The tea, which contained more water-soluble principles, was more effective in this regard than the alcoholic extract. Some constipation, probably due to the tannin, was also noted. The investigators concluded that Mormon tea does not belong to the exceedingly active class of medicinal plants and that the properties usually attributed to it are already "well supplied by some well-established therapeutic agent."[9]

This is a sound comment. If you enjoy the astringent flavor of Mormon tea and are not concerned about its high tannin content, you will be satisfied. If you expect it to have any pronounced therapeutic effect, you will be disappointed.

REFERENCES

1. D. W. Stevenson. Ephedraceae. In *Flora of North America North of Mexico,* Volume 2. Flora of North America Editorial Committee (Eds.). Oxford University Press, New York, 1993, pp. 428-434.
2. R. E. Terry. *Journal of the American Pharmaceutical Association* 16: 397-400, 1927.
3. W. H. Hylton (Ed). *The Rodale Herb Book.* Rodale Press Book Div., Emmaus, Pennsylvania, 1976, pp. 513-514.
4. D. G. Spoerke Jr. *Herbal Medications.* Woodbridge Press Publishing Co., Santa Barbara, California, 1980, pp. 122-123.
5. A. Gottlieb. *Legal Highs.* 20th Century Alchemist, Manhattan Beach, California, 1973, p. 37.
6. D. B. Mowrey. *Herbalist* 2(6): 26, 1977.
7. M. Castleman. *The Healing Herbs.* Rodale Press, Emmaus, Pennsylvania, 1991, pp. 158-161.
8. R. Hegnauer. *Chemotaxonomie der Pflanzen,* Volume 1. Birkhäuser Verlag, Basel, Switzerland, 1962, pp. 460-462.
9. R. E. Terry, op. cit., pp. 397-400, 1927.

Muira Puama

Botanists have had almost as much trouble identifying the source of muira puama as zoologists have experienced in locating the Loch Ness monster. Although the existence of the plant (unlike that of the monster) is beyond dispute, it was once thought to derive from *Liriosma ovata* Miers or, perhaps, *Acanthea virilis* (nomen nudum). Finally, scientists identified it as the stem-wood and root of two Brazilian shrubs, *Ptychopetalum olacoides* Benth. and *Ptychopetalum uncinatum* Anselmino. Both are members of the family Olacaceae.[1] However, all of the aforementioned species are offered in the herb trade as muira puama.

Also known as potency wood, the drug has a long history of use in Brazil as a powerful aphrodisiac and nerve stimulant. It is an ingredient in a number of proprietary remedies and folk medicines for sexual impotence. Muira puama is also touted for dyspepsia, menstrual irregularities, rheumatism, and paralysis caused by poliomyelitis and as a general tonic and appetite stimulant.[2]

The drug is administered by mouth, either as a powder, an alcoholic extract, or a decoction (extract formed by boiling in water). An alternative method of obtaining the aphrodisiac effect is to bathe the genitals with a concentrated decoction. It is also applied locally to treat rheumatism and muscle paralysis.

Chemical studies show that muira puama contains as its principal constituent 0.4 to 0.5 percent of a mixture of esters, two-thirds of which is behenic acid, lupeol, and β-sitosterol; in the remaining portion, other fatty acids replace the behenic acid.[3] Other more-or-less routine plant constituents, such as volatile oil, resin, fat, tannin, various fatty acids, and the like, have been isolated from muira puama.

None of the constituents in this drug is known to exhibit any pronounced physiological activity. This, plus the lack of any reported clinical testing of muira puama, causes us to view its reported effects with considerable skepticism. Until such tests have been carried out, no claims of efficacy or safety can be substantiated, and we must conclude at this point that potency wood is impotent instead.

REFERENCES

1. V. E. Tyler, L. R. Brady, and J. E. Robbers. *Pharmacognosy,* Ninth Edition. Lea and Febiger, Philadelphia, Pennsylvania, 1988, p. 482.

2. E. F. Steinmetz. *Quarterly Journal of Crude Drug Research* 11: 1787-1789, 1971.

3. H. Auterhoff and E. Pankow. *Archiv der Pharmazie* 301: 481-489, 1968.

Mullein

Verbascum thapsus L., the common mullein of the United States, is a woolly biennial herb belonging to the family Scrophulariaceae. During the first year, its large, hairy leaves form a low-lying rosette. In the spring of the second year, a tall stem develops from the leaves to a height of four feet or more and is topped by a spike of yellow flowers. Both the leaves and flowers of this and of closely related *Verbascum* species have been used in folk medicine. The flowers are particularly popular in Europe and are usually obtained from *V. phlomoides* L. or *V. thapsiforme* Schrad., species native to that continent.[1]

According to Grieve, mullein is valuable in the treatment of such a wide range of ailments as to make the newest "wonder drug" seem inactive in comparison.[2] It is believed to possess demulcent, emollient, and astringent properties and is useful in treating both bleeding of the lungs (tuberculosis) and of the bowels. Not only is it both a sedative and a narcotic, but it can also be useful in treating cases of asthma, coughs, and hemorrhoids. Burns and erysipelas (streptococcus infections) yield to its application, as do bruises, frostbite, diarrhea, ear infections, most disease germs, and migraine. As if this were not enough—it is also useful in driving away evil spirits. This "wonder" drug is taken internally, applied locally, and even smoked, to treat these various conditions. Some of the more practical but less medicinal applications of mullein include using the yellow flowers as a blond hair dye and wearing the fuzzy leaves in the stockings to keep the feet warm.

Returning now to reality, a number of different chemical constituents, including a mucilage, saponins, and tannins, have been identified in both mullein leaves and flowers.[3] None of these compounds in the quantities present possesses any important therapeutic activity, although they do have some mild demulcent (soothing), expectorant, and astringent properties. Because the amount of mucilage in the plant is relatively small (2 percent), the herb's utility in cough preparations is probably due more to the expectorant action of the contained saponins than to the soothing effects of the mucilage.[4] Antiviral activity has been reported in laboratory studies (against influenza and herpes simplex viruses), along with mild expectorant activity. Mullein flowers seem to be preferred over the leaves in European phytomedicine. The German regulatory authorities allow use of the flowers as a soothing expectorant in catarrh of the upper respiratory tract.[5]

Recently seven new saikosaponin homologues, deemed mulleinsaponins I through VII, have been isolated from *V. thapsiforme* and a number of other Asian *Verbascum* species.[6] A new triglycoside of luteolin, verbascoside, extracted from the whole plant of *V. thapsus*, has been reported.[7]

In Europe, the flowers are a common ingredient in many of the popular herbal mixtures sold as medicinal teas for their palliative effects in minor coughs and colds. In the United States, both the leaves and flowers formerly enjoyed official status in *The National Formulary* but were deleted in 1936 since it was believed that, except for its soothing mucilage, the drug lacked therapeutic virtue.[8]

REFERENCES

1. H. W. Youngken. *Textbook of Pharmacognosy*, Sixth Edition. The Blakiston Co., Philadelphia, Pennsylvania, 1948 p. 800.

2. M. Grieve: *A Modern Herbal,* Vol. 2. Dover Publications, New York, 1971, pp. 562-566.

3. P. H. List and L. Hörhammer (Eds.). *Hagers Handbuch der Pharmazeutischen Praxis.* Fourth Edition, Volume 6C. Springer-Verlag, Berlin, 1979, pp. 417-422.

4. J. Kraus and G. Franz. *Deutsche Apotheker Zeitung* 127: 665-669, 1987.

5. *The Lawrence Review of Natural Products,* September, 1989.

6. T. Miyase, C. Horikoshi, S. Yabe, S. Miyaska, F. R. Melek, and G. Kusana: *Chemical and Pharmaceutical Bulletin* (Tokyo) 45: 2029-2033, 1997.

7. R. Mehrotra, B. Ahmed, R. A. Vishwakarma, and R. S. Thakur. *Journal of Natural Products* 52: 640-643, 1989.

8. A. Osol and G. E. Farrar Jr. *The Dispensatory of the United States of America,* Twenty-Fourth Edition. J. B. Lippincott, Philadelphia, Pennsylvania, 1947, p. 1644.

Myrrh

. . . and looking up they saw a caravan of Ishmaelites coming from Gilead, with their camels bearing gum, balm, and myrrh, on their way to carry it down to Egypt.

Genesis 37:25

Such quotations make us aware that myrrh has been an important article of commerce from ancient times, but it was valued then as a constituent of incense and perfumes, as well as one of the main ingredients in the embalming process, rather than as a medicinal agent. Technically, myrrh is an oleo-gum-resin (a mixture of volatile oil, gum, and resin) obtained from *Commiphora myrrha* (Nees) Engl., *Commiphora molmol* Engl. (Somalian myrrh), *Commiphora mada,gascariensis* Jacq. [Abyssian myrrh; syn. *C. abyssinica* (Berg) Engl.] or other species of *Commiphora.* These are small trees of the family Burseraceae, native to Ethiopia, Somalia, and the Arabian peninsula. Myrrh consists of irregular masses or tear-shaped pieces, dark yellow or reddish brown in color, that exude naturally or from incisions made in the bark. The different commercial varieties are named according to their source, for example, Somali myrrh and Arabian myrrh.[1]

Modern herbalists recommend myrrh as an antiseptic. It is incorporated into a salve that is applied externally in treating hemorrhoids, bed sores, and wounds. The tincture (alcoholic solution) is considered an effective oral astringent and is used as a mouthwash or for treating sore throat and similar conditions. Myrrh is taken internally for indigestion, ulcers, and to relieve bronchial congestion. It even enjoys some reputation as an emmenagogue (stimulates menstrual flow).[2,3] The suggestion that it can be therapeutic in cancer, leprosy, and syphilis is farfetched.[4]

Myrrh contains about 8 percent of a volatile oil, 25 to 40 percent of resin, and about 60 percent of gum. Various aldehydes and phenolic constituents in the volatile oil combine with acidic constituents in the resin to produce some astringent and antiseptic properties in the oleo-gum-resin. The physical properties of the gum and resin also confer a protective action on the mixture. Although myrrh is presently an ingredient in several commercial mouthwashes, it is far more widely used as a fragrance component in soaps, cosmetics, and perfumes and a flavor component in food products such as candy, baked goods, and so on. German regulatory authorities allow myrrh powder and tincture for the topical treatment of mild inflammations of the oral and pharyngeal mucosa.[5]

Our ancestors, who valued the oleo-gum-resin primarily for its fragrance, were perhaps better informed about the appropriate use of myrrh than we are. As a botanical, it is apparently relatively nontoxic but possesses only mild astringent and protective properties.

REFERENCES

1. V. E. Tyler, L. R. Brady, and J. E. Robbers. *Pharmacognosy,* Ninth Edition. Lea and Febiger, Philadelphia, Pennsylvania, 1988, p. 151.

2. M. Tierra. *The Way of Herbs.* United Press, Santa Cruz, California, 1980, pp. 105-106.

3. R. Lucas. *Nature's Medicines.* Wilshire Book Co., North Hollywood, California, 1977, pp. 71-75.

4. A. Y. Leung and S. Foster. *Encyclopedia of Common Natural Ingredients Used in Food, Drugs, and Cosmetics,* Second Edition. John Wiley and Sons, New York, 1996, pp. 382-383.

5. *Bundesanzeiger,* October 15, 1987.

Nettle

One would think that a high-technology society capable of splitting the atom and sending a man to the moon would long ago have learned everything there is to know about the stinging nettle. Not so. Even the agent responsible for the skin irritation produced by contact with the leaves of this common plant remains nearly as much a mystery to twentieth-century scientists as it did to the first caveman who stumbled against it. Its erect stalk, two to three feet in height, bears dark green leaves with serrated margins and small, inconspicuous flowers. Botanists now designate it *Urtica dioica* L. and place it in the family Urticaceae.

The American material differs from the typical European *Urtica dioica* subspecies *dioica* primarily in that it has male and female flowers on the same plant. Some botanists treat the varieties of *U. dioica* subspecies *gracilis* as separate species. The four *Urtica* species (with two subspecies and six varieties) that occur in North America have stinging hairs.[1] (After accidental contact with it, people usually refer to the nettle by various uncomplimentary titles.)

The entire plant, collected just before flowering, has had a lengthy reputation in folk medicine as a specific for asthma. It has also been given as an expectorant, antispasmodic, diuretic, astringent, and tonic.[2] Applying nettle to the scalp, especially in the form of the fresh juice, was said to stimulate hair growth. Cases of chronic rheumatism have been treated by placing nettle leaves directly on the afflicted area. Roman soldiers, facing the inhospitable climate in Britain, reportedly used the same irritation produced by nettle leaves to keep their legs warm.[3] The tender tops of young, first-growth nettles are believed especially palatable when cooked; Gibbons gives a number of recipes that use them, including nettle pudding and nettle beer.[4]

Numerous analyses of nettle have revealed the presence of more than twenty different chemical constituents;[5] few of them would provide any pronounced therapeutic activity from the plant when taken internally. Although the local irritation produced by the stinging hairs is real enough, there is just no evidence to show that it is effective in treating rheumatism or growing hair on bald heads. The principles in the hairs thought to be responsible for this irritant action include histamine, acetylcholine, and 5-hydroxytryptamine. However, studies on plants of the closely related, but more toxic, genus *Laportea* have cast doubt on this, and the identity of the compound responsible for the pain from contact with nettle remains to be established.[6]

Many ancient and modern herbalists assert that rubbing fresh dock (Rumex) leaves on the affected area will reduce the stinging discomfort of nettle rash. There is even an old rhyme:

> Nettle in, dock out.
> Dock rub nettle out!

No objective evidence supports this claim aside from the fact that firm rubbing—by itself—was found to produce a short-lived lessening of the pain inflicted by *Laportea* species.[7] It is also possible that the time and effort spent on finding a dock leaf is sufficient to distract the victim from the itching caused by nettle rash.[8]

Nettle is rich in chlorophyll and serves as a readily available commercial source of that pigment. Young nettle shoots are edible when cooked and contain approximately the same amounts of carotene (provitamin A) and vitamin C as spinach or other similar greens. The diuretic properties of nettle leaf have long been recognized, and several pharmaceutical preparations incorporating it are currently marketed in Europe for this purpose. In addition, an extract of nettle root has become quite popular there in recent years for the treatment of urinary retention brought on by benign prostatic hypertrophy (enlargement of the prostate gland not due to cancer). Some clinical evidence attests to its effectiveness, including eight open and observational studies and two placebo-controlled, double-blind studies.[9] Therefore, German health authorities now allow it to be used for this condition.[10,11] Additional studies are needed to verify this or any other traditional medical use of nettle.

REFERENCES

1. S. Foster and R. Caras. *A Field Guide to Venomous Animals and Poisonous Plants.* Houghton Mifflin Co., Boston, Massachusetts, 1994, pp. 138.

2. M. Grieve. *A Modern Herbal,* Volume 2. Dover Publications, New York, 1971, pp. 574-579.

3. R. C. Wren and R. W. Wren. *Potter's New Cyclopaedia of Botanical Drugs and Preparations,* New Edition. Health Science Press, Hengiscote England, 1975, p. 216.

4. E. Gibbons. *Stalking the Healthful Herbs,* Field Guide Edition. David McKay Co., New York, 1970, pp. 133-138.

5. W. Holzner (Ed.). *Das Kritische Heilpflanzen Handbuch.* Verlag ORAC, Vienna, 1985, p. 18.

6. W. V. MacFarlane. *Economic Botany* 17: 303-311, 1963.

7. W. V. MacFarlane. In *Venomous and Poisonous Animals and Noxious Plants of the Pacific Region,* H. L. Keegan and W. F. MacFarlane (Eds.). Macmillan, New York, 1963, pp. 31-37.

8. J. Mitchell and A. Rook. *Botanical Dermatology.* Greengrass Ltd., Vancouver, BC, Canada, 1979, p. 38.

9. V. Schulz, R. Hänsel, and V. E. Tyler. *Rational Phytotherapy: A Physicians' Guide to Herbal Medicine,* Third Edition. Springer, Berlin, 1998, pp. 228-229.

10. P. Goetz. *Zeitschrift für Phytotherapie* 10: 175-178, 1989.

11. *Bundesanzeiger,* January 5, 1989; March 6, 1990.

New Zealand Green-Lipped Mussel

An extract prepared from the New Zealand green-lipped mussel is just one more "therapeutic substance" in a long line of such nondrugs advocated for the relief of symptoms of arthritis. The mussel, known technically as *Perna canaliculus* a member of the family Mytilidae, is cultivated on special marine farms; at an unspecified time in its growth cycle, it is extracted by an undesignated process to yield a product of largely unknown composition, which is stabilized by freeze-drying.[1] It is commonly marketed in the form of 625 mg capsules.

Advocates maintain that daily doses of the extract relieve the discomfort of both rheumatoid and osteoarthritis in patients of any age, without any dangerous side effects. They recommend taking the product for periods of three to eighteen weeks and cite relief of pain, restoration of mobility, and reduction of joint distortion, even in very elderly people who have been crippled for years with the disease. Mussel extract is also claimed to produce a feeling of well-being and a desire to be active. Sixty percent of persons treated are supposed to have shown improvement.

Croft indicates that how mussel extract works is unknown, although he does note that it appears to work directly "on the origin of the inflammation rather than on the inflammation itself."[2] He further reports that although the extract contains amino acids and minerals, all attempts to fractionate it to obtain and identify an active principle have resulted in a total loss of activity. Long contradicts these statements, pointing out that the mussel extract contains mucopolysaccharides of the hyaluronic-acid type, which act by increasing the viscosity of the joint lubricants in the body, thereby relieving swelling and stiffness. They also are believed to contribute to the extract's effectiveness in healing soft tissues, torn tendons, and bruises.[3] Yet there is no evidence to support the latter assertions. Although such mucopolysaccharides are important in forming the cement that binds body cells together and in producing other gel-like materials,[4] when taken by mouth, they would be largely broken down by the digestive process prior to absorption and thus be unable to carry out these functions.

Scientific studies in small animals, as well as clinical investigations in human beings, have not produced any substantial evidence of the effectiveness of mussel extract for arthritis. No anti-inflammatory action could be demonstrated in rats when the extract was given to them orally. A six-month clinical trial carried out on patients in a Scottish hospital seemed to have some positive results, but the "clinical data were poorly controlled."[5] To date, several human trials have not been able to reproduce the initial positive results for anti-inflammatory activity in the treatment of symptoms of rheumatoid arthritis and osteoarthritis first reported in the early 1980s.[6]

About the best that can be said for New Zealand green-lipped mussel extract for now is that, except in persons allergic to shellfish, it does not seem to have any pronounced toxicity nor to produce any serious side effects. It certainly has equally little in the way of therapeutic benefits.

REFERENCES

1. J. E. Croft. *Relief from Arthritis: A Safe and Effective Treatment from the Ocean.* Thorsons Publishers Limited, Wellingborough, England, 1979.
2. Ibid.
3. M. L. Long. *Herbalist New Health* 6(4): 20, 1981.
4. C. H. Best and N. B. Taylor. *The Physiological Basis of Medical Practice,* Seventh Edition. Williams and Wilkins, Baltimore, Maryland, 1961, p. 21.
5. Anon. *Lancet* I: 85, 1981.
6. *Lawrence Review of Natural Products,* April 1997.

Pangamic Acid ("Vitamin B_{15}")

"When I use a word," Humpty Dumpty said in a rather scornful tone, "it means just what I choose it to mean—neither more nor less."

Lewis Carroll
Through the Looking Glass

Pangamic acid (variously known as vitamin B_{15}, pangamate, calcium pangamate, Russian formula, etc.) is one of those words, referred to by Humpty Dumpty, whose meaning varies according to its user's intent. As far as identity is concerned, it may mean almost anything; as far as therapeutic utility is concerned, it means absolutely nothing.

Originally, pangamic acid was the name given to a compound which Ernst Krebs Sr. and Ernst Krebs Jr. claimed to have isolated from the kernels of apricots (*Prunus armeniaca* L.) in 1943 and which was subsequently trade named by them as "vitamin B_{15}." Remember that Krebs Jr. was the discoverer of laetrile, a product also obtained from apricot kernels. Pangamic acid was identified as D-gluconodimethyl aminoacetic acid, an ester of D-gluconic acid and dimethylglycine. A U.S. patent, issued in 1949, claimed that pangamate was able to detoxify toxic products formed in the human system. It was also supposed to be effective in treating asthma and allied diseases, conditions of the skin and the respiratory tract, painful nerve and joint afflictions, cell proliferation (cancer?), eczema, arthritis, neuritis, and the like. No scientific or clinical evidence was presented in support of these claims.[1]

At the present time, a number of companies market pangamic acid, but the true identity of the product is known only to Humpty Dumpty because it varies with the producer. Pangamic acid consists variously of one or more of the following ingredients, often simply mixed together: sodium gluconate, calcium gluconate, glycine, diisopropylamine dichloroacetate, dimethylglycine, calcium chloride, dicalcium phosphate, stearic acid, Avicel (a form of cellulose), and so on.[2] Since there is absolutely no standard of chemical identity for products sold under the name pangamic acid (or any of its various synonyms), one authority takes the position that the preparation has no real existence.[3]

Unfortunately, the commercial tablets sold in various "health food" outlets under this name, and the exorbitant price charged for such products (about $20.00 for 100 tablets), are all too real. Besides, depending on its composition, pangamic acid may be toxic to human beings. Gluconic acid, glycine, and acetate are relatively inert, but isopropylamine acts on smooth muscles to lower blood pressure. Yet many of the advertisements for pangamic acid claim it improves oxygenation of the heart, brain, and other vital organs.[4]

Dichloroacetate may cause oxalic acid stones and other kidney problems; it has also been shown to be mutagenic by the Ames test and therefore has the potential to induce cancer. Dimethylglycine may react with nitrites in the intestine to form dimethylnitrosamine, a potent carcinogen. Calcium chloride also has poisonous properties.[5]

None of these constituents of the various pangamic acid products has been shown to have any nutritional or medical utility. Consequently, early in 1981, the FDA increased its seizures of the product after legal cases in several district courts had upheld the agency's actions in similar test cases. Unfortunately, the number of FDA seizures has been relatively small in comparison to the enormous amount of this totally worthless and potentially dangerous product offered for sale.[6,7] In the 1990s, pangamic acid continues to be marketed throughout the United States, both as such and in the form of "equivalents," such as dimethylglycine.

REFERENCES

1. V. Herbert. *Nutrition Cultism: Facts and Fiction.* George F. Stickley Co., Philadelphia, Pennsylvania, 1980, pp. 107-120.

2. R. J. Moleski (Ed.). *Informer* (University of Rhode Island, College of Pharmacy, Drug Information Service) 3(6): 1-4, 1979.

3. V. Herbert, op. cit., pp. 107-120.

4. J. B. Laudano. *Recipe (*St. John's University, College of Pharmacy and Allied Health Professions) 16(2): 7-8, 1979.
5. V. Herbert, op. cit., pp. 107-120.
6. T. H. Jukes. *Journal of the American Medical Association* 242: 719-720, 1979.
7. Anon. *Drug Topics* 125(5): 20, 1981.

Papaya

The papaya plant is a small tree, *Carica papaya* L. (family Caricaceae), native to tropical America but found in tropical areas throughout the world. Its trunk, which is nonwoody and hollow, produces large, deeply lobed leaves and smooth-skinned cantaloupe-like fruits or melons directly on its surface without intervening branches. When ripe, the fruits are a very desirable food. Shallow cuts made on the surface of fully grown but unripe fruits cause them to exude a milky sap or latex that after collection and drying is known as crude papain. In addition to the large quantities produced by incising the fruit, about 2 percent of papain is found in papaya leaves.[1]

Papain, or vegetable pepsin as it is sometimes called, is a mixture of proteolytic enzymes with a fairly broad spectrum of activity; it hydrolyzes not only proteins but small peptides, amides, and some esters as well. Other components of the crude enzyme mixture hydrolyze both carbohydrates and fats.[2] This wide range of activity accounts for the use of papain in folk medicine for digestive disturbances of all kinds but particularly for those associated with protein-rich foods. The enzyme and the papaya leaves are also employed as a vermifuge (expels intestinal worms), especially for tapeworms.[3] As a digestive aid, papaya tablets containing between 10 and 50 mg of papain are commercially available.

Face creams, lotions, cleansers, and so on are often formulated with papain in the belief that the enzyme will exert "a digestive effect on freckles and other sun blemishes" while cleansing the pores of makeup and providing a general "softening" effect. However, the use of papain most familiar to every housewife is as a meat tenderizer. The enzyme mixed with salt as an activator and a carbohydrate dispersing agent is sold in every supermarket. When shaken on tough meat before cooking, especially beef, it acts as an effective tenderizer by predigesting to some degree the fibrous animal protein. Various commercial applications of papain such as chill-proofing beer and clarifying fruit juices are interesting but beyond the scope of this discussion.[4]

Those who drink a tea prepared from papaya leaves as a digestive aid should be aware that (according to French) the leaves should first have been subjected to a fermentation process similar to that used for black tea. This is said to facilitate extraction of the active principles by boiling water and to brew a much richer beverage than is obtained with ordinary dried papaya leaves.[5] Unfortunately, papain is quite unstable in the presence of digestive juices, so its efficacy as a vermifuge or digestive aid is open to serious question.[6] After examining the evidence supporting the supposed effectiveness of papaya, German health authorities have concluded that its utility remains unproven, and its therapeutic use is not recommended.[7]

The 1978 report by Indian investigators that papain was teratogenic (produced birth defects) and embryotoxic (poisonous to the embryo) in rats needs verification before it can be given much credence. A more realistic concern is the enzyme's ability to induce allergic responses in sensitive individuals.[8]

REFERENCES

1. C. D. French. *Papaya: The Melon of Health.* Arco Publishing Co., New York, 1972.
2. V. E. Tyler, L. R. Brady, and J. E. Robbers. *Pharmacognosy,* Ninth Edition. Lea and Febiger, Philadelphia, Pennsylvania, 1988, p. 276.
3. M. Pahlow. *Das Grosse Buch der Heilpflanzen.* Gräfe und Unzer GmbH, Munich, 1979, p. 405.
4. C. D. French, op. cit.
5. C. D. French, op. cit.
6. H. Wagner. *Pharmazeutische Biologie: Drogen und Ihre Inhaltsstoffe,* Second Edition. Gustav Fischer Verlag, Stuttgart, Germany, 1982, pp. 306-307.
7. *Bundesanzeiger,* August 18, 1987.
8. A. Y. Leung and S. Foster. *Encyclopedia of Common Natural Ingredients Used in Food, Drugs, and Cosmetics,* Second Edition. John Wiley and Sons, New York, 1996, pp. 402-405.

Parsley

Parsley is like the weather: everyone knows about it, but no one does anything about it. Parsley is certainly our most familiar herb, widely employed as a culinary garnish for more than 2,000 years, but it is seldom eaten.

The leaf, root, and fruit of *Petroselinum crispum* (Mid.) Nym. (family Apiaceae) have also been used for centuries in folk medicine. Botanists indicate that the plant's leaves are pinnate decompound, which simply means that they are divided and somewhat featherlike in their appearance. Since parsley can be identified by anyone who ever ate in a restaurant, here are the essentials: It is a widely cultivated, biennial herb with yellow flowers borne in clusters. Its fruits, commonly called seeds, are small, ovate, and grayish to grayish brown with alternating ribs and furrows.[1]

In classical medicine, parsley fruits were used primarily as a stomachic or carminative (aids digestion and expels gas), and the root as a diuretic (increases flow of urine). The fruit also enjoyed some reputation as an emmenagogue and an abortifacient (stimulates menstrual flow and abortion).[2] Although there may be some basis in fact for these uses of parsley, such attributes as a cure for diabetes, heart problems, liver ailments, and venereal disease are purely fanciful.[3]

Using parsley as a digestive aid, diuretic, and emmenagogue is based on its volatile oil content, the concentration of which varies from less than 0.1 percent in the root, to about 0.3 percent in the leaf, and from 2 to 7 percent in the fruit. As is the case with many plants that have been cultivated for centuries, many varieties of parsley exist. The chemical composition of the volatile oil obtained from some of these varieties is quite variable. So-called German parsley oil contains 60 to 80 percent of apiol (parsley camphor) as its principal component; French parsley oil contains less apiol but more (50 to 60 percent) myristicin, a compound originally found in nutmeg oil but very similar to apiol, both chemically and in its physiological action.[4] Both apiol and myristicin are uterine stimulants, accounting for the use of parsley volatile oil as an emmenagogue and for its misuse as an abortifacient.

Although it is not commonly eaten in quantity, parsley herb is a good natural source of carotene (provitamin A), vitamins B_1, B_2, and C, as well as iron and other minerals. It is therefore a good nutrient, especially when combined with bulgur and other ingredients in the tasty Lebanese salad, tabbouleh, but as a drug, the herb is of little worth.

Because of their relatively high content of volatile oil, the fruits (seeds) may possess some stomachic and diuretic properties, but both such actions are relatively mild. Parsley volatile oil with its contained apiol and myristicin is toxic, and under no circumstances should it be administered to pregnant women.[5] Since efficacy of parsley fruits is not well documented, and risks outweigh benefits, the German health authorities do not recommend their use.[6] So, while the parsley plant is of little medicinal value, its volatile oil is literally:

> A remedy too strong for the disease.

Sophocles
Tereus
Fragment II. 589

REFERENCES

1. H. W. Youngken. *Textbook of Pharmacognosy,* Sixth Edition. The Blakiston Co., Philadelphia, Pennsylvania, 1948, pp. 632-633.
2. V. E. Tyler, L. R. Brady, and J. E. Robbers. *Pharmacognosy,* Ninth Edition. Lea and Febiger, Philadelphia, Pennsylvania, 1988, p. 483.
3. M. S. Keller. *Mysterious Herbs and Roots.* Peace Press, Culver City, California, 1978, pp. 264-279.
4. H. A. Hoppe. *Drogenkunde,* Eighth Edition, Volume 1. Walter de Gruyter, Berlin, 1975, pp. 817-818.
5. T. Sollmann. *A Manual of Pharmacology,* Seventh Edition. W. B. Saunders, Philadelphia, Pennsylvania, 1948, p. 148.
6. *Bundesanzeiger,* March 2, 1989.

Passion Flower

The passion flower derives its name from the imagined resemblance of its floral parts to the elements surrounding the crucifixion of Christ. Its three styles represent the three nails, its ovary looks like a hammer, the corona is the crown of thorns, the ten petals represent the ten *true* apostles (excluding Peter who denied Him and, of course, Judas), and so on. The Moldenkes remind us that the plant was unknown in biblical times, and the fancied symbolism dates from 1610.[1]

Passiflora incarnata L. is a fast-growing perennial vine (family Passifloraceae) occurring from Virginia to southern Illinois and southeast Kansas, south to Florida and Texas. It is known by the names passionflower, maypop, and apricot vine.[2]

Passion flower was introduced into medicine in 1840 by Dr. L. Phares of Mississippi. The remedy remained buried in obscurity until Prof. I. J. M. Goss of Atlanta, Georgia, reintroduced it into the practice of eclectic physicians in the late nineteenth century.[3] For many years, the dried flowering and fruiting top of the perennial climbing vine *Passiflora incarnata* has enjoyed a reputation as a calmative agent and sedative. It was listed in *The National Formulary* from 1916 to 1936 but has since fallen into disuse in the United States. Without valid evidence to support taking passion flower extract as a sedative or nighttime sleep aid,[4] the FDA has not recognized it as generally safe or effective since 1978. Still, it continues to be incorporated into many sedative-hypnotic drug mixtures marketed in Europe. A sedative chewing gum containing passiflora extract and vitamins was patented in 1978 in Romania.[5]

Constituents responsible for the pharmacological activity of passion flower have been the subject of ongoing research throughout most of this century. The plant does contain one or more so-called harmala alkaloids, but their number and identity are disputed. Besides, such alkaloids generally act as stimulants, not depressants. A Polish report that both an alkaloid fraction and a flavonoid pigment fraction produced sedative effects in mice was subsequently followed up by Japanese investigators.[6] They were able to isolate small amounts of the pyrone derivative maltol from an alkaloid-containing extract of the plant. Maltol was found to induce depression in mice and to exhibit other sedative properties. The scientists concluded that the depressant effects of maltol, no doubt, counteracted the stimulant action of the harmala alkaloids but were not strong enough to explain the total sedative effects of the plant extract. Flavonoids including vitexin, isovitexin, isoorientin, schaftoside, and isoschaftoside have also been identified and may contribute to biological activity.[7] Further studies are obviously necessary before the active principles of passion flower can definitely be identified.

Reports in the literature that passion flower contains toxic, cyanogenic glycosides are misleading. Spoerke, for example, makes such a statement but has confused the medicinally used passion flower, *Passiflora incarnata,* with the commonly cultivated, ornamental blue passion flower, *Passiflora caerulea* L.[8] The latter species does contain cyanogenic glycosides, but the plant we have been discussing does not.[9] The genus *Passiflora*, containing both edible and potentially toxic species, is represented by only two temperate North American species. The remaining 430 species occur in the tropical Americas, with twenty species in Indomalaysia.[10]

Even though passion flower is not recognized as a safe or effective drug in the United States, it is an ingredient in many pharmaceutical products sold in Europe as sedatives. German health authorities note that reduced activity has been reported in animals to which the drug was administered and have approved its use for conditions involving nervous unrest.[11]

REFERENCES

1. H. N. Moldenke and A. L. Moldenke. *Plants of the Bible.* Chronica Botanica Co., Waltham, Massachusetts, 1952, p. xiv.

2. E. P. Killip. *Contributions from the U.S. National Herbarium* 351: 1-23, 1960.

3. H. W. Felter and J. U. Lloyd. *King's American Dispensatory,* Eighteenth Edition, Volume 2. The Ohio Valley Co., Cincinnati, Ohio, 1900, pp. 1439-1441.

4. *Federal Register* 43(114): 25578, June 13, 1978.

5. F. Gagiu, T. Budiu, P. Lavu, and O. Bidiu. Romanian Patent No. 59, 589. In *Chemical Abstracts* 89: 48897n, 1978.

6. N. Aoyagi, R. Kimura, and T. Murata. *Chemical and Pharmaceutical Bulletin* 22: 1008-1013, 1974.

7. ESCOP. *Proposals for European Monographs on the Medicinal Uses of* Passiflorae Herba. European Scientific Cooperative for Phytotherapy, Meppel, The Netherlands, 1992.

8. D. G. Spoerke Jr. *Herbal Medications.* Woodbridge Press Publishing Co., Santa Barbara, California, 1980, pp. 134-135.

9. R. Hegnauer. *Chemotaxonomie der Pflanzen*, Volume 5. Birkhäuser Verlag Basel, Switzerland, 1969, p. 295.

10. D. J. Mabberley. *The Plant Book,* Second Edition. Cambridge University Press, Cambridge, United Kingdom, 1997, p. 532.

11. *Bundesanzeiger,* November 30, 1985; March 6, 1990.

Pau d'Arco

Perhaps the most popular herbal cancer "cure" that has appeared in recent times is pau d'arco tea, also known as ipe roxo, lapacho, or taheebo tea. This beverage is prepared from the bark of various species of *Tabebuia,* a genus of about 100 broad-leaved, mostly evergreen trees of the family Bignoniaceae, native to the West Indies and Central and South America. Referred to in Brazil as ipe or pau d'arco, these plants have an extremely hard wood that is most attractive and practically indestructible.[1] Its resistance to decay probably attracted the attention of the natives to the medicinal potential of the species.

Popular reports state that Indian tribal doctors in Brazil brew a tea from the inner bark of *Tabebuia avellanedae* or *Tabebuia altissima,* known respectively as lapacho colorado and lapacho morado, that is used to treat cancer as well as ulcers, diabetes, and rheumatism. Proponents also claim that pau d'arco is "a powerful tonic and blood builder" and is effective against rheumatism, cystitis, prostatitis, bronchitis, gastritis, ulcers, liver ailments, asthma, gonorrhea, ringworm, and even hernias.[2,3] The drug is claimed to have been popular in the old Inca Empire, long before the Spanish invaded the New World. *T. avellanedae* is native to the warmer parts of South America, but *T. altissima* supposedly grows high in the Andes Mountains, where "not even the worst winter storms can blow it down."

Such popular reporting leaves much to be desired. There is no plant with the scientific name *Tabebuia altissima;* further, no species of *Tabebuia* grows high in the Andes. This remote habitat was apparently the creation of some advertising copywriter to make the drug sound more exotic. Although *Tabebuia avellanedae* is a name found in literature, the correct botanical designation of the species is *Tabebuia impetiginosa* (Mart.) Standl.

Complicating the matter of origin even further is the fact that some pau d'arco herbal teas marketed in this country do not derive from the *Tabebuia* species at all, even though they are labeled as lapacho colorado or lapacho morado. Instead, they are stated to represent the bark of *Tecoma curialis* Solhanha da Gama, another closely related member of the same plant family. This probably makes little difference because the useful constituents and therapeutic activities, if any, are undoubtedly similar. It nevertheless leaves the botanical source of pau d'arco products unclear. The outstanding American botanical authority on this group of plants, the late Dr. A. H. Gentry, speculated that probably all of the bark in question was being obtained from some lowland *Tabebuia* species.[4]

Because of their commercial significance in the construction industry, *Tabebuia* woods have been examined in detail. In addition to such therapeutically uninteresting constituents as volatile oils, resins, bitter principles, and the like, they contain from 2 to 7 percent of a naphthoquinone derivative known as lapachol. Analyses of the barks of three *Tabebuia* species showed that, unlike the wood, they did not contain lapachol and dehydro-α-lapachone as their major naphthoquinone constituents. Depending on the species, these compounds were either present in traces or entirely absent. Three lapachol derivatives were instead detected; their physiological properties appear to be very similar to those of lapachol.[5]

Lapachol does possess some anticancer properties. In 1968, it was shown to have significant activity against Walker 256 carcinosarcoma, particularly when administered orally to animals in which this tumor had been implanted. In later studies, lapachol was found to be active against other kinds of animal cancers, including Yoshida sarcoma and Murphy-Sturm lymphosarcoma. In trials with human cancer patients, however, as soon as effective plasma levels were attained, undesirable side effects were severe enough to require that the drug be stopped. These included moderate to severe nausea, vomiting, anemia, and a tendency to bleed.[6] Animal and other laboratory studies have demonstrated that lapachol also possesses antibiotic, antimalarial, and antischistosomal properties, but scientific studies have not been done in humans because of the problem of toxicity.

Pau d'arco is marketed in the United States as a tea or "dietary supplement," with no therapeutic claims made on product labels. Its lack of proven effectiveness, its potential toxicity, and its relatively high cost render its use both unwise and extravagant.

REFERENCES

1. W. B. Mors and C. T. Rizzini. *Useful Plants of Brazil.* Holden-Day, Inc., San Francisco, California, 1966, p. 125.

2. A. de Montmorency. *Spotlight,* January 5 and 12, 1981, pp. 10, 11, 34; June 8, 1981, p. 6.

3. B. Wead. *Second Opinion: Lapacho and the Cancer Controversy.* Rostrum Communications, Vancouver, BC, Canada, 1985, 196 pp.

4. A. H. Gentry. Personal communication, September 9, 1983.

5. M. Girard, D. Kindack, B. A. Dawson, J.-C. Ethier, D. V. C. Awang, and A. H. Gentry. *Journal of Natural Products* 51: 1023-1024, 1988.

6. J. B. Block, A. A. Serpick, W. Miller, and P. H. Wiernik. *Cancer Chemotherapy Reports* (Part 2) 4(4): 27-28, 1974.

Pennyroyal

Two very different members of the mint family (Lamiaceae) are referred to as pennyroyal. Although different in appearance, *Hedeoma pulegioides* (L.) Pers., the American pennyroyal, and *Mentha pulegium* L., European or Old World pennyroyal, possess similar chemical compositions and applications. A tea prepared from the leaves of either pennyroyal has been recommended as a stimulant, carminative, diaphoretic, and emmenagogue.[1] An emmenagogue promotes the menstrual flow but in popular writing is often a euphemism for an abortifacient. The more active pennyroyal oil has been taken in attempted abortion with tragic results.

American pennyroyal contains up to 2 percent of a volatile oil, and European pennyroyal up to 1 percent of an even more disagreeable-smelling volatile oil.[2] Both oils consist of 85 to 92 percent pulegone and are therefore quite toxic, causing severe liver damage even in relatively small amounts. Two tablespoonfuls of pennyroyal oil caused the death of an eighteen-year-old expectant mother despite intensive hospital treatment initiated just two hours after she took it.[3]

Two cases of multiple organ failure in infants have also been reported. In the case of an eight-week-old male infant, severe liver damage with cerebral edema and necrosis resulted in death after four days. In a second case, a week later in the same hospital, similar organ failure was reported in a six-month-old male infant who recovered after two months of hospitalization. In both cases, the mothers gave the infant boys pennyroyal tea.[4] Clearly, the minute amounts of essential oil in the tea can be dangerous to infants. As little as ½ teaspoonful of the oil produced convulsions and coma in one individual.[5] At least twenty-four cases of pennyroyal toxicity, mostly from ingestion of small amounts of the oil in adult women, have been reported in the medical literature.[6]

While pennyroyal oil may indeed induce abortion, it does so only in lethal or near-lethal doses. Such amounts would ordinarily not be obtained from drinking a tea prepared from the herb. Nevertheless, popular belief has it that pennyroyal tea is an effective abortifacient, as indicated by this description of the low upbringing of a virtueless woman:

> . . . she was the fifth of twelve children in the river-bottom family, with a mother who laid the cards and brewed tansy, pennyroyal and like concoctions for luckless girls who were in need.

Mari Sandoz
*Slogum House**

Pennyroyal tea possesses no therapeutic properties that could not be obtained from more pleasant and effective medicaments. Therefore, internal consumption of the herb has little to recommend it. Both the herb and the oil are commonly employed in collars or powders used to keep cats and dogs free from fleas.[7] Bunches of pennyroyal plants hung up to dry are also said to be effective mosquito repellants.[8]

REFERENCES

1. N. Coon. *Using Plants for Healing,* Second Edition. Rodale Press, Emmaus, Pennsylvania, 1979, p.119.
2. E. Guenther. *The Essential Oils,* Volume 3. D. Van Nostrand, New York, 1949, pp. 575-586.
3. J. B. Sullivan Jr., B. H. Rumack, H. Thomas Jr., R. G. Peterson, and P. Bryson. *Journal of the American Medical Association* 242: 2873-2874, 1979.
4. J. A. Bakerink, S. M Gospe Jr., R. J. Diman, and M. W. Eldridge. *Pediatrics* 98: 944-947, 1996.
5. E. F. Early. *Lancet* II: 580-581, 1961.
6. I. B. Anderson, W. H. Mullen, J. E. Meeker, S. C. Khojastech-Bakht, S. Oishi, S. D. Nelson, and P. D. Blanc. *Annals of Internal Medicine* 124: 726-734, 1996.
7. C. Wittmann. *Herb Quarterly* 48: 12-16, 1990.
8. V. E. Tyler. *Hoosier Home Remedies.* Purdue University Press, West Lafayette, Indiana, 1985, p. 98.

---

*Copyright 1937, 1965 by Mari Sandoz. Published by the University of Nebraska Press. (By permission.)

Peppermint

Because it is such a popular flavoring agent and so widely used in just about every kind of product intended for human consumption, you would think peppermint was one of our oldest herbs. But it is not. The plant is a natural hybrid or cross that sprouted in a field of spearmint growing in England in 1696. Ever since that time, peppermint, *Mentha* x *piperita* L. of the family Lamiaceae, has been intensively cultivated for its fragrant volatile oil. Since it does not breed true from seed, peppermint is vegetatively propagated; there are numerous cultivated varieties.[1]

Peppermint is used primarily for its stimulating, stomachic, and carminative properties in treating indigestion, flatulence (gas), and colic.[2] It is usually taken in a moderately warm tea prepared from the leaves, several cups being slowly sipped to bring fairly prompt relief. In Europe, the aromatic herb is incorporated in many tea mixtures intended to alleviate various ailments of the stomach, intestines, and liver. Although it may well contribute to certain actions of these mixtures, it is often used simply as a pleasant flavor.

As an aid to digestion, its activity is due primarily to its contained volatile oil, which exists in the herb (leaves and flowering tops) in concentrations ranging from 1 to 3 percent. American peppermint oil contains from 50 to 78 percent of free menthol and another 5 to 20 percent of various combined forms (esters) of menthol.[3] These major components are also largely responsible for peppermint's ability to stimulate the bile flow and promote digestion along with certain other flavonoid pigments with similar properties.

In addition, the volatile oil acts as a spasmolytic, reducing the tonus of the lower esophageal (cardial) sphincter and facilitating eructation (belching).[4] This antispasmodic property may also account for the popularity of peppermint tea as a household remedy for menstrual cramps. Peppermint oil temporarily inhibits hunger pangs in the stomach, but soon the stomach resumes its peristaltic movements, which then become stronger than before.[5] In this way it works to stimulate the appetite.

Peppermint is generally recognized as being safe for human consumption.[6] However, the Food and Drug Administration has recently declared the oil to be ineffective as a digestive aid and banned its use as a nonprescription drug for that purpose in this country. This does not mean that the drug is necessarily ineffective. It simply means that the FDA was not presented with evidence proving its efficacy. For financial reasons relative to the high cost of a new drug application, that would not be feasible in this country. In contrast to the FDA's declaration, the more realistic German health authorities have found both peppermint and its volatile oil to be effective spasmolytics—especially for conditions in the upper digestive tract—antibacterial agents, and promoters of gastric secretions.[7]

There is a tendency among laypeople, and even some herbalists,[8] to confuse the two most popular mints, peppermint and spearmint, because the plants are similar in appearance, and both are used as flavoring agents. In other respects they are very different. Peppermint owes its therapeutic utility to the presence of menthol, which is not present in spearmint.[9] Consequently, the principal use of spearmint is as a flavor, not as a digestive aid.

Adults find peppermint tea to be a pleasant-tasting beverage and a useful remedy for mild digestive disturbances and related complaints. However, the tea should not be given to infants or very young children, since they may often experience a choking sensation from the contained menthol.

REFERENCES

1. S. Foster. *Peppermint,* Mentha *x* piperita, *Botanical Series No. 306*, Second Edition. American Botanical Council, Austin, Texas, 1996, 7 pp.

2. M. Grieve. *A Modern Herbal,* Volume 2. Dover Publications, New York, 1971, pp. 537-543.

3. V. E. Tyler, L. R. Brady, and J. E. Robbers. *Pharmacognosy,* Ninth Edition. Lea and Febiger, Philadelphia, Pennsylvania, 1988, pp. 113-118.

4. P. H. List and L. Hörhammer (Eds.). *Hagers Handbuch der Pharmazeutischen Praxis,* Fourth Edition, Volume 5. Springer-Verlag, Berlin, 1976, pp. 767-771.

5. T. Sollmann. *A Manual of Pharmacology,* Seventh Edition. W. B. Saunders, Philadelphia, Pennsylvania, 1948, p. 167.

6. *Lawrence Review of Natural Products,* July 1990.

7. *Bundesanzeiger,* November 30, 1985; March 13, 1986.

8. M. Castleman. *The Healing Herbs.* Rodale Press, Emmaus, Pennsylvania, 1991, pp. 253-256.

9. R. Hänsel and H. Haas. *Therapie mit Phytopharmaka.* Springer-Verlag, Berlin, 1983, p. 180.

Pokeroot

Pokeroot comes from *Phytolacca americana* L. (family Phytolaccaceae), a large, much-branched, perennial herb that bears rather spectacular clusters of dark purple, almost black, berries. The plant is a wayside weed from New England to Texas. If there is any ailment for which pokeroot has not been recommended, it is simply because the herbalists have not yet thought of it. It is variously described as an alterative, cathartic, emetic, a narcotic, and a gargle, as well as a remedy for conjunctivitis, cancer, dyspepsia, glandular swelling, chronic rheumatism, ringworm, scabies, and ulcers.[1]

Pokeroot is not therapeutically useful for anything. It may act as an emetic and cathartic, but it does so because it is extremely toxic, due to the presence of a saponin mixture called phytolaccatoxin. The plant also contains a proteinaceous mitogen, PWM, that may produce various abnormalities of the blood cells after absorption.[2]

Children have died and adults have been hospitalized from the gastroenteritis, hypotension, and diminished respiration caused by eating pokeroot or the leaves of the plant. With the exception of the ripe berries, all parts of the mature plant are considered very poisonous. Some controversy exists about the relative toxicity of the berries, but they are nevertheless widely consumed as a folk remedy for rheumatism and arthritis. The very young shoots, which some use as potherbs, are believed to be innocuous. However, it doesn't seem prudent to eat them when so many other safer sources of greens are available.

A forty-three-year-old Wisconsin woman drank one cup of tea prepared from ½ teaspoonful of the root and required twenty-four hours of intense hospital treatment before her condition stabilized. As a result of this case, the Herb Trade Association issued a policy statement declaring that pokeroot should not be sold as an herbal food or beverage. It further recommended that all packages containing it carry an appropriate warning statement regarding the plant's toxicity and the potential danger if taken internally.[3] Despite this recommendation, modern herbalists, such as Santillo, continue to recommend the use of pokeroot for a wide variety of conditions ranging, alphabetically, from cirrhosis of the liver to swollen breasts.[4]

Eating the first shoots of the spring as a green, usually boiled in two changes of water, is still a common practice in the South, where bundles of poke leaves are still widely available at farmers markets. Ingesting the leaves as a potherb sometimes results in poisoning, characterized by severe gastroenteritis, with intense vomiting and frothy diarrhea. At least four cases of cardiotoxicity have also been reported following ingestion of the roots and/or leaves.[5]

With or without a warning label, pokeroot is definitely not recommended for either internal or external use by human beings. In the words of Lewis Carroll's Mad Gardener:

> "Were I to swallow this," he said, "I should be very ill!"

REFERENCES

1. D. I. Macht. *Journal of the American Pharmaceutical Association* 26: 594-599, 1937.

2. W. H. Lewis and P. R. Smith. *Journal of the American Medical Association* 242: 2759-2760, 1979.

3. Anon. *Whole Foods* 2(4): 14, 1979.

4. H. Santillo. *Natural Healing with Herbs,* Eighth Printing. Hohm Press, Prescott, Arizona, 1990, pp. 162-163.

5. R. J. Hamilton, R. D. Shih, and R. S. Hoffman. *Veterinary and Human Toxicology* 37: 66-67, 1995.

Pollen

Exotic, even bizarre, remedies, ranging from peacock excrement to moss grown on the skull of a man who had died by violence, have long been part of our medical lore.[1] The use of pollen does have historical roots. Ragweed pollen is said to have been sprinkled onto the faces of captured Aztec enemies before their sacrifice.[2] In fact, man has been extremely diligent in searching out such unusual materials, possibly in the hope that they may possess unusual curative properties. Pollen is a relatively recent example of such a drug. Although pollen extracts have been used for many years to detect and provide immunity against allergies, it is only during the past few years that pollen itself has become widely available in the form of tablets, capsules, extracts, and the like, which are recommended for a variety of ailments.

Pollen consists of microspores (male reproductive elements) of seed-bearing plants. Often the marketed product is designated bee pollen, implying that a mixture of pollens from various plants was collected by honeybees. Indeed, a meshlike pollen trap has been developed that relieves bees of a portion of the pollen carried on their back legs as they reenter the hive. But there is no way to determine if a particular pollen grain was originally collected by a bee or not, so it seems best to refer to the commercially available material simply as pollen.

Enthusiasts declare that pollen will either provide relief for or cure such conditions as premature aging, cerebral hemorrhage, bodily weakness, anemia, weight loss, enteritis, colitis, and constipation.[3] Pollen has also been promoted for use in weight loss regimens, indigestion, neurasthenia, and brain damage.[4] It is also touted as having general tonic properties—promoting better health along with happiness and optimism. Studies conducted in Sweden and Japan seem to indicate the drug may be of value in treating chronic prostatism. An Austrian report found pollen useful in alleviating the symptoms of radiation sickness in patients being treated for cancer of the cervix.[5]

The chemical constituents of pollen have been rather extensively investigated. Although the different components vary greatly in quantity among pollens of different species, some general ranges may be quoted. Polysaccharides, particularly starch and cell-wall constituents, constitute up to 50 percent of a typical pollen. Low molecular weight carbohydrates (simple sugars) make up another 4 to 10 percent. The concentration of lipids (fats, oils, and waxes) is extremely variable, ranging from 1 to 20 percent. Protein exists to the extent of 5.9 to 28.3 percent, but only 0.5 to 1.0 percent of the total protein is allergenic in nature. About 6 percent of free amino acids are also present. Other constituents include about 0.2 percent of carotenoid and flavonoid pigments plus small amounts of terpenes and sterols. Some pollens are quite high in vitamin C; concentrations ranging from 3.6 to 5.9 percent have been reported.[6] In addition, pollen components are affected by the season of the year, harvest methods, and of course, the species from which it is derived.

None of the identified constituents of pollen has been linked to any significant therapeutic activity as advocated by its enthusiasts. The few studies in which favorable results were obtained require repetition and reevaluation before being accepted as factual. In the meantime, keep in mind that many pollens can induce severe allergic responses when inhaled or eaten, at least in some individuals. Birch pollen has been associated with cross-reactive allergy to apples, carrots, and celery tuber (celeriac).[7]

Pollen's continuing appeal to an uncritical public was demonstrated when a snack bar containing it was chosen as the "official snack food" of the 1987 Pan American Games in Indianapolis. After the bee pollen in the bars was found to contain ragweed pollen, a serious, perennial allergen for many Indiana residents, the product was widely denounced by nutritionists and FDA officials. The producer claimed the bars provided an "energy boost" lasting for several hours; authorities noted that this effect was due to the honey and other carbohydrates, not to the small amount of pollen, contained in the snack.[8]

Since pollen has no significant therapeutic or nutritive value that cannot be obtained more easily and cheaply from other sources, it cannot be recommended for either purpose. And since its allergenic properties may render it downright hazardous to some, we must actively discourage its use both as a medicine and as a food.

REFERENCES

1. A. C. Wootton. *Chronicles of Pharmacy,* Reprint Edition, Volume 2. USV Pharmaceutical Corp., Tuckahoe, New York, 1972, pp. 2-3.

2. H. F. Linskens and W. Jorde. *Economic Botany* 51: 78-87, 1997.

3. G. J. Binding. *About Pollen.* Thorsons Publishers Ltd., Wellingborough, England, 1971.

4. H. F. Linskens and W. Jorde, op. cit., p. 82.

5. Anon. *Bee Pollen—A Short Treatise.* Les Ruchers de la Côte d'Azur, New York, 1977.

6. R. G. Stanley and H. F. Linskens. *Pollen.* Springer-Verlag, New York, 1974.

7. H. F. Linskens and W. Jorde, op. cit., p. 84.

8. L. G. Calceca. *The Indianapolis Star,* December 8, 1986, pp. 15-16.

Propolis

Unlike pollen, of relatively recent medicinal use, propolis, or bee glue, was an official drug in the London pharmacopeias of the seventeenth century.[1] However, there was a long hiatus in its popularity between the seventeenth and the late twentieth century; now propolis once again is receiving considerable attention from laypersons and scientists both. The unusual drug is a brownish resinous material collected by bees from the buds of various poplar and conifer trees and used by the insects to fill cracks or gaps in their hives.

Those who advocate its therapeutic use claim that propolis has an antibacterial activity greater than that of penicillin and other common antibiotic drugs.[2] They maintain the product "works" by raising the body's natural resistance to infection through stimulation of the immune system. It is supposed to be especially beneficial in the treatment of tuberculosis. Duodenal ulcers and gastric disturbances are also thought to benefit from propolis therapy. Applied externally in the form of a cream, advocates say it relieves various types of dermatitis, especially those caused by bacteria and fungi.[3] Propolis is commercially available in the form of capsules (both pure and combined with 50 percent pollen), throat lozenges, cream, chips (used like chewing gum), and as a powder (to prepare a tincture).

More than twenty-five different constituents of propolis have now been tested scientifically against various species of bacteria and fungi for antibacterial and antifungal effects. Results indicate that the antimicrobial properties of the drug are attributable mainly to the flavonoids pinocembrin, galangin, pinobanksin, and pinobanksin-3-acetate; in addition, *p*-coumaric acid benzyl ester and a caffeic acid ester mixture were also active. Pinocembrin, a 5,7-dihydroxyflavanone, showed considerable antifungal activity. However, none of these isolated principles was as effective as various antibiotics or sulfa drugs with which they were compared: streptomycin, oxytetracycline, chloramphenicol, nystatin, griseofulvin, and sulfamerazine.[4] A series of studies on propolis carried out by Polish investigators showed that besides bacteriostatic and fungistatic properties, the drug inhibited the growth of protozoa, accelerated bone formation, had regenerative effects on tissues, stimulated some enzyme actions, and showed cytostatic effects (inhibited cell growth and division).[5] Recent studies have shown mild anti-inflammatory activity in vitro along with weak antioxidant (free-radical scavenging) activity.[6] Two studies have shown that ethanolic extracts of Cuban red propolis (perhaps a name coined in anticipation of future entry of commercial products in world markets) have shown potential protective effects against allyl-alcohol- and carbon tetrachloride-induced liver toxicity in rats.[7,8] It must be emphasized that all of these results were obtained from experiments carried out in vitro, that is, in the chemical laboratory outside the living body, or in small animals. Double-blind clinical trials in human beings have apparently never been conducted with propolis.

The flavonoid pigments of propolis seem to possess modest antibacterial and antifungal properties but much less active than the standard drugs for controlling such microorganisms. A recent Italian study found that antimicrobial activity against gram-positive bacteria, yeast and dermatophytes, showed that solvents used for various commercial propolis preparations played an important role in antimicrobial activity. Edible oil (type not specified), propylene glycol, and ethanol solutions maintained inhibition for more than two weeks, while glycerin extract produced some inhibition for only two days.[9]

Other tentative claims for potential therapeutic utility require clinical verification. In the interim, it is safe—and appropriate—to continue using propolis to seal openings in bee hives, for which it has proven highly effective.

REFERENCES

1. A. C. Wootton. *Chronicles of Pharmacy,* Reprint Edition, Volume 2. USV Pharmaceutical Corp., Tuckahoe, New York, 1972, p. 2.

2. S. S. Jones. *Whole Foods* 3(9): 26-30, 1980.

3. T. Smith and S. S. Jones. *Herbalist* 4(11): 12-13, 1979.

4. J. Metzner, H. Bekemeier, M. Paintz, and E. Schneidewind. *Die Pharmazie* 34: 97-102, 1979.

5. B. Hladón, W. Bylka, M. Ellasin-Wojtaszek, L. Skrzypcza, P. Szafarek, A. Chodera, and Z. Kowalewski. *Arzneimittel-Forschung* 30: 1847-1848, 1980.
6. *Lawrence Review of Natural Products,* February 1996.
7. R. Gonzalez, I. Corcho, D. Remirez, S. Rodriguez, O. Ancheta, N. Merino, A. Gonzalez, and C. Pascual. *Phytotherapy Research* 9: 114-117, 1995.
8. D. Remirez, R. Gonzalez, S. Rodriguez, O. Ancheta, J. C. Bracho, A. Rosado, E. Rojas, and M. E. Ramos. *Phytomedicine* 4: 309-314, 1997.
9. B. Tosi, A. Donini, C. Romagnoli, and A. Bruni. *Phytotherapy Research* 10: 335-336, 1996.

Pygeum

Pygeum, as represented in the herb market, is the bark of an African tree of the family Rosaceae, *Prunus africana* (Hook. f) Kalkman, whose common name derives from the now obsolete botanical designation *Pygeum africanum* Hook. f. The tree is present in highland mountain forests in Africa and Madagascar, occurring in Afromontane forest "islands" from 4,500 to 6,000 feet. Surrounding forests have been clear-cut for forest products and agricultural land, limiting the tree's habitat. The bark harvest, primarily taken from the wild in Cameroon, Kenya, Tanzania, Madagascar, as well as the Democratic Republic of Congo (the former Zaire), has had a devastating effect on wild populations of the species.[1,2] This overexploitation sparked conservation concerns, resulting in the species being listed in Appendix II of CITES (Convention on International Trade in Endangered Species of Wild Fauna and Flora) in order to monitor species in international trade.[3]

The fresh bark, leaf, and fruits contain amygalin, yielding hydrocyanic acid when crushed; hence, they have an almond flavor. In Africa, the fresh leaves have been mixed with milk to produce a subsitute for almond milk.[4] In African counties the bark is used by traditional healers for inflammation, kidney disease, malaria, stomachache, and fever, among other uses. In Natal, the bark is infused in milk and used to treat problems of difficult urination. In Cameroon, the bark has been used to treat fever and madness and has a local reputation as an aphrodisiac. The root and bark have traditionally been used in southern, eastern, and central Africa for inflammation of the prostate gland and kidney disease.[5]

Folkloric use in Africa attracted the attention of European researchers, and a patent was issued in 1966 for use of a pygeum bark extract in the treatment of benign prostatic hyperplasia (BPH). The bark contains pentacyclic terpenes, including ursolic, oleanolic, and crataegolic acids, plus *n*-docosanol and *n*-teracosanol as described active constituents. Phytosterols present, including beta-sitosterol, beta-sitosterone, and campesterol, may contribute to biological activity. Pygeum products are standardized to contain 14 percent triterpenes and 0.5 percent *n*-docosanol.[6]

Similar to saw palmetto berry and stinging nettle root, the bark of pygeum is valued in European phytotherapy for the treatment of benign prostatic hypertrophy. Primary clinical experience for the extract has centered in France and Italy, rather than Germany, where saw palmetto extracts dominate the market for BPH phytotherapy. Pharmacological studies have shown that the extract possesses anti-inflammatory activity (by inhibiting enzymes involved in depolymerization of proteoglycans in prostate connective tissue), reduces cholesterol levels in the prostate (limiting androgen synthesis), and inhibits prostaglandin synthesis. It has also been shown to increase prostatic secretions in both rats and humans, along with improving the composition of seminal fluid.[7]

In the past two decades, twenty-six clinical trials have been conducted on pygeum extracts, at a dose of 100 to 200 mg per day, half of which were double-blinded versus placebo. The results indicate positive effects in the treatment of symptoms associated with BPH such as difficulty in urination, frequent nighttime urge to urinate, and a reduction of residual urine volume. Transient side effects involving gastrointestinal irritation (inducing nausea and abdominal pain) have been reported in clinical trials.[8]

Since BPH is not self-limiting or self-diagnosable, pygeum should be used under medical supervision. Given the environmental impact of the bark harvest, consumers are advised to consider carefully the consequences of choosing pygeum products.

REFERENCES

1. M. Cunningham, A. B. Cunningham, and U. Shippmann. *Trade in* Prunus africana *and the Implementation of CITES.* Germany Federal Agency for Nature Conservation, Bonn, 1997.

2. A. B. Cunningham and F. T. Mbenkum. *Sustainability of Harvesting* Prunus africana *bark in Cameroon.* UNESCO, Paris, 1993.

3. S. Foster. *101 Medicinal Herbs.* Interweave Press, Loveland, Colorado, 1998.

4. J. M. Watt and M. G. Breyer-Brandwijk. *Medicinal and Poisonous Plants of Southern and Eastern Africa,* Second Edition. E & S Livingstone Ltd, Edinburgh, Scotland, 1962, p. 894.

5. T. Sutherland and N. Ndam. *Medicinal Plants of the Limbe Botanic Garden.* Limbe Botanic Garden, Cameroon, pp. 54-55, nd.

6. *Lawrence Review of Natural Products,* January 1998.

7. Ibid.

8. V. Schulz, R. Hänsel, and V. E. Tyler. *Rational Phytotherapy: A Physicians' Guide to Herbal Medicine,* Third Edition. Springer, Berlin, 1998, pp. 232-233.

Raspberry

Although the flavorful fruits of the red raspberry, varieties of *Rubus idaeus* L. and *R. strigosus* Michx. (family Rosaceae), were once used rather extensively to give a pleasant taste to various pharmaceutical preparations, it is the leaves of the plants that are still used as a popular folk remedy. The former species is a native of Europe, the latter of North America. Both prickly stemmed shrubs are now widely cultivated in the United States.

Raspberry leaves are employed for their astringent and stimulant properties.[1] Supporters recommend a strong infusion (tea) as a gargle or mouthwash for sore mouth and inflammation of the mucous membrane of the throat, as well as for various wounds and ulcers when applied locally to them. The moistened leaves may also be applied externally as a poultice. Drinking cold raspberry leaf tea as a remedy for diarrhea is said to give immediate relief to that and to various stomach ailments.

Incidentally, with one exception, the leaves of blackberry (*Rubus fruticosus* L.) are used in a similar fashion to raspberry leaves.[2] All of the comments about raspberry's astringent properties apply to both drugs. With respect to raspberry leaves, however, it must be noted that tea made from them has acquired a considerable reputation as "the drink" for expectant mothers. In the popular literature, the beverage is praised as a "panacea during pregnancy which is said to do everything from allaying morning sickness to preventing miscarriage to erasing labor pains."[3] Even a reputable scientific reference credits it as a traditional remedy for painful and profuse menstruation and for use before and during confinement.[4]

The raspberry leaves' astringent properties for treating sore mouth or diarrhea, etc., are readily explained by an appreciable content of hydrolyzable tannin, containing both gallic and ellagic acids in the free and combined forms.[5] An extract of the leaves has been reported to have little effect on uterine muscles of laboratory animals but, in pregnant rats, inhibited contraction of those muscles. Fractions, with as yet unidentified compounds, have been shown to stimulate smooth muscle action, especially of the uterine muscle. Another fraction was found that reduced uterine contractions.[6] These are very preliminary studies that raise far more questions than answers. Without adequate clinical studies, it is impossible to say if the drug's reputation as a relaxant of the smooth muscles of the uterus and intestine is real or imagined.

As with most green leafy plant materials, fresh raspberry leaves contain quantities of vitamin C. How much is present in the commercially available dried leaves depends on the conditions of drying and the manner and time of storage.

At present, we lack sufficient evidence to support any outstanding therapeutic importance of raspberry. It does have an astringent action and might be called upon occasionally as a modestly effective mouthwash and gargle or diarrhea treatment. If pregnant women believe that it provides relief from various unpleasant effects associated with their condition, no harm is done because it is relatively inexpensive and, as far as we now know, relatively harmless except for its tannin content.

Incidentally, if you wish to purchase raspberry tea for its supposedly beneficial effects, be sure to read the label carefully and make certain that raspberry leaves are the principal ingredient. Many so-called raspberry teas are simply ordinary black teas flavored with a volatile oil that smells like raspberry fruits. Raspberry leaves do not have this characteristic fruity aroma. In my personal opinion, the best thing about raspberry is its fresh fruit, which I still like on my breakfast cereal.

REFERENCES

1. M. Grieve. *A Modern Herbal,* Volume 2. Dover Publications, New York, 1971, pp. 671-672.

2. M. Pahlow. *Das Grosse Buch der Heilpflanzen.* Gräfe und Unzer GmbH, Munich, Germany, 1979, pp. 164-166.

3. T. Clifford. *Cures.* Macmillan Publishing Co., Inc., New York, 1980, p. 52.

4. J. E. F. Reynolds, Ed: Martindale: *The Extra Pharmacopoeia,* Twenty-Ninth Edition. The Pharmaceutical Press, London, 1982, p. 1609.

5. P. H. List and L. Hörhammer (Eds.). *Hagers Handbuch der Pharmazeutischen Praxis,* Fourth Edition, Volume 6B. Springer-Verlag, Berlin, 1979, pp. 186-188.

6. C. J. Briggs and K. Briggs. *Canadian Pharmaceutical Journal* 41-43, April 1997.

Red Bush Tea

It is a pleasure, for a change, to discuss a plant material that has no therapeutic value and for which none is claimed. Indeed, red bush or rooibos tea is valued not only for its taste but also because it is devoid of any undesirable physiological effects. Although there has been much confusion about the proper name of the plant whose dried leaves and fine twigs constitute the tea, it is now generally agreed to be *Aspalathus linearis* (Burm. f.) R. Dahlgr., also sometimes designated *Borbonia pinifolia* Marloth or *Aspalathus contaminata* (Thunb.) Druce.[1]

This member of the family Fabaceae is a shrub native to the mountainous regions of South Africa. It attains a height of six feet and bears long needlelike leaves that turn a brick red color when bruised. The plant is now extensively cultivated, particularly in the Cedarberg Mountains near the Clanwilliam district. In late summer or early autumn, the plants are harvested, cut into short lengths, moistened, bruised, allowed to ferment, and dried in the sun to produce red bush tea, which is brewed like ordinary tea to make South Africa's most popular hot beverage.[2]

Red bush tea is drunk either plain or with sugar and milk according to the consumer's taste. It is valued highly not only for its refreshing flavor—a liking for it may be acquired—but because it is low in tannin (less than 5 percent) and is essentially caffeine-free. These factors combine with the presence of some vitamin C in the tea (0.0016 percent) to produce a beverage that is quite acceptable to those who wish to avoid caffeine and high tannin concentrations in their hot drinks. Red bush tea is currently marketed in the United States under the trade name Kaffree Tea.

REFERENCES

1. R. Dahlgren. *Botaniska Notiser* 117: 188-196, 1964.
2. R. H. Cheney and E. Scholtz. *Economic Botany* 17: 186-194, 1963.

Red Clover

During the early years of the twentieth century, more than a half dozen major pharmaceutical companies manufactured and marketed various "Trifolium Compound" preparations. The formula of one extract produced by the Wm. S. Merrell Chemical Company of Cincinnati, Ohio, included red clover, the blossoms of *Trifolium pratense* L. (family Fabaceae), plus seven other vegetable drugs. This extract and the fluidextracts and syrups of other manufacturers were widely sold as alteratives, that is, cures for venereal disease.[1] As early as 1912, the Council on Pharmacy and Chemistry of the American Medical Association reported, "We have no information to indicate that they (red clover preparations) possess medicinal properties."[2] Still, trifolium continued to be listed in *The National Formulary* until 1946.

A 1991 catalog of "health products" lists Red Clover Combination formula in which red clover is combined with four of the seven herbs contained in the previously discredited mixture, plus a few additional ingredients, added primarily for flavor.[3] As is customary in such publications, no specific therapeutic use is described, but the $7.95 price tag indicates the product should be good for something. Reference to a typical modern herbal will of course reveal that red clover is an alterative.[4] All this is reminiscent of the expression (attributed to Marie Antoinette's milliner), "There is nothing new except what is forgotten." Unfortunately, in this case, what had been forgotten should have remained forgotten.

The characteristic red blossoms of this extensively cultivated forage plant have been subjected to detailed chemical analyses. More than one-third of a page of fine print in a recent reference is required just to list the names of the chemical compounds detected in red clover.[5] Yet none of these various pigments, phenolic compounds, tannins, and the like has any pronounced therapeutic value, particularly in the treatment of venereal disease. The statement that red clover tea sweetened with honey and drunk two or three times a day for a period of four to six weeks will purify the blood (euphemism for cure venereal disease) is simply not factual.[6] It is true that the obvious symptoms of the disease may disappear in that time, but no cure has been obtained.

Four isoflavones found in red clover blossoms, formononetin, biochanin A, daidzein and genistein, do have mild estrogenic activity. This perceived effect has resulted in the introduction of a widely advertised red clover blossom dietary supplement product from Australia as a "natural choice for maintaining estrogen." Clinical studies to support these claims are apparently under way.[7]

The isoflavones found in red clover, and more common food sources such as soy beans, may alter hormone production or metabolism, intracellular enzymes, cell differentiation and production, and growth factors.[8] Epidemiological studies suggest potential value of isoflavones in chemoprevention (cancer prevention), in Asian countries where soy products, rich in isoflavones, are widely consumed. A recent in vitro study found that biochanin A from red clover inhibited carcinogen activation in cell cultures, suggesting the need for further studies.[9]

Potential for future benefits pending further research is just that. This emphasizes one of the real dangers of self-medicating with ineffectual drugs. They themselves may not be harmful, but neglecting effective treatment for a serious disease may eventually prove disastrous. Exactly the same precaution applies to the local application of red clover flowers to treat "cancerous growths."[10] This herb is also an ingredient in the "internal formula" of the now discredited Hoxsey cancer treatment.[11] Red clover is *not* effective in such conditions, and delay in obtaining proper therapy may be fatal.

REFERENCES

1. H. W. Felter and J. U. Lloyd. *King's American Dispensatory,* Eighteenth Edition, Volume. 2. The Ohio Valley Co., Cincinnati, Ohio, 1900, pp. 1995-1996.

2. *A Reprint of the Reports of the Council on Pharmacy and Chemistry of the American Medical Association with the Comments that Appeared in the Journal During 1912.* American Medical Association, Chicago, Illinois, 1913, p. 40.

3. Indiana Botanic Gardens, Hammond, Indiana, Spring 1991, p. 28.

4. M. Grieve. *A Modern Herbal,* Volume 1. Dover Publications, New York, 1971, pp. 207-208.

5. P. H. List and L. Hörhammer (Eds.). *Hagers Handbuch der Pharmazeutischen Praxis,* Fourth Edition, Volume 6C. Springer-Verlag, Berlin, 1979, pp. 265-266.

6. M. Pahlow. *Das Grosse Buch der Heilpflanzen.* Gräfe und Unzer GmbH, Munich, Germany, 1979, pp. 352-353.

7. Anon. Press Release, Novogen, Inc., April 7, 1998.

8. D. Pathak, K. Pathak, and A. K. Singla. *Fitoterapia* 62: 371-389, 1992.

9. J. M. Cassady, T. M. Zennie, Y. H. Chae, M. A. Ferin, N. E. Portuondo, and W. M. Baird. *Cancer Research* 48: 6257-6261, 1988.

10. M. Grieve. *A Modern Herbal,* Volume 1. Dover Publications, New York, 1971, pp. 207-208.

11. J. A. Lowell. *Nutrition Forum* 4: 89-91, 1987.

Rose Hips

Because of their relatively high content of vitamin C, the bright scarlet to deep red, ovoid or pear-shaped fruits or hips of several species of roses always occupy a significant place in discussions of natural medicines. Most commonly, the hips are collected from the dog rose *Rosa canina* L., but the larger hips of the Japanese rose, *R. rugosa* Thunb., are valued highly, as are those of *R. acicularis* Lindl. and *R. cinnamomea* L. All are more or less familiar members of the family Rosaceae.[1]

Rose hips are used to prepare teas, extracts, purees, marmalades, even soups, all of which are consumed for their vitamin C content.[2] The extracts are also incorporated into a number of "natural" vitamin preparations, including tablets, capsules, syrups, and the like. Most such preparations are careful not to state on the label exactly how much of the vitamin C content is derived from rose hips and how much from synthetic ascorbic acid. In addition to their antiscorbutic (antiscurvy) properties, rose hips have a mild laxative and slight diuretic action.

Although fresh rose hips contain concentrations of vitamin C ranging from 0.5 to 1.7 percent, the actual content of the commercially available dried fruit is extremely variable depending on the exact botanic source, where it was grown, when it was collected, how it was dried, and when and where it was stored, etc. Indeed, many commercial samples of the plant material no longer contain detectable amounts of vitamin C. Even if we assume that they contain an average of 1 percent of the vitamin and that all of the vitamin is present in the finished preparation—two propositions that are not necessarily valid—the present cost of vitamin C from rose hips is about twenty-five times more than the synthetic product.[3]

Rose hips contain, in addition to vitamin C, a large number of different chemical compounds, including about 11 percent of pectin and 3 percent of a mixture of malic and citric acids. These are probably responsible for the mild laxative and diuretic effects of the drug.[4]

Based solely on cost, one must reject rose hips as an economical source of vitamin C. Of course, if you are able to collect and process your own, or if you simply like the taste of rose hip tea or similar preparations, that is quite a different matter. If you are still interested in the commercial product for its vitamin content—in spite of the cost—I would not recommend purchase unless the material is clearly labeled as to the amount of natural vitamin C actually contained in it. This is really asking very little. Relatively simple assay procedures exist and any well-equipped food or drug quality-control laboratory is capable of conducting them.[5] Only in this way can you be assured of value received for any money you spend on rose hips.

REFERENCES

1. N. Coon. *Using Plants for Healing,* Second Edition. Rodale Press, Emmaus, Pennsylvania, 1979, p. 174.

2. J. C. Torke. *Herbalist* 1: 472-473, 1976.

3. V. E. Tyler, L. R. Brady, and J. E. Robbers. *Pharmacognosy,* Ninth Edition. Lea and Febiger, Philadelphia, Pennsylvania, 1988, p. 485.

4. P. H. List and L Hörhammer (Eds.). *Hagers Handbuch der Pharmazeutischen Praxis,* Fourth Edition, Volume 6B. Springer-Verlag, Berlin, 1979, pp. 164-170.

5. P. H. List and L Hörhammer (Eds.), Ibid.

Rosemary

While it may be useful in the culinary arts, rosemary is not one of the most valuable herbs from the medicinal viewpoint. That it is one of the best known is attested to by frequent references, some occurring even in children's literature.

> Old Mrs. Rabbit was a widow; she earned her living by knitting. . . . She also sold herbs and rosemary tea . . .
>
> Beatrix Potter
> *The Tale of Benjamin Bunny*

Rosemary consists of the leaves or the leaves with flowering tops of *Rosmarinus officinalis* L. (family Lamiaceae), an evergreen shrubby herb with aromatic linear leaves, which are dark green above and white below, and small pale blue flowers. It has been extensively cultivated in so-called kitchen gardens. If members of the women's liberation movement were to seek a plant to represent their cause, it would certainly be rosemary, for there is an old English belief that the plant will thrive only in the garden of a household where the "mistress" is really the "master."[1]

Various preparations of rosemary, including an infusion or tea, a wine, a spirit (alcoholic solution), and a bath, are recommended for their tonic, astringent, and diaphoretic (increases perspiration) effects. The leaves are also said to have stomachic (aids digestion) properties and to make a hair tonic that, when applied externally, will prevent baldness.[2] Rosemary is recommended especially in cases of low blood pressure; a bath prepared from it is so stimulating to the body that it should not be taken in the evening or it may prevent one from sleeping.[3] Finally, both the drug and its volatile oil have been used as emmenagogues (to stimulate menstrual flow) and abortifacients.

Whatever physiological activity rosemary possesses is attributed to its volatile oil, which occurs in the leaves in concentrations ranging from 1 to 2.5 percent.[4] Containing such compounds as camphor, borneol, and cineole, the volatile oil, similar to many others, has antibacterial properties. It also has some stimulating properties, particularly when applied locally. The leaves of rosemary contain a number of flavonoid pigments, one of which, diosmin, is reported to decrease capillary permeability and fragility.[5] German health authorities have approved its use internally for indigestion and as a supportive therapy for rheumatic disorders; externally, it is recognized for the treatment of circulatory disturbances.[6] However, the exact extent of rosemary's therapeutic usefulness remains unknown, for no systematic clinical studies with the leaves have been reported.

Rosemary is extensively used as a household spice and as a flavoring agent in various commercial products including prepared meats, baked goods, vegetables, and so on. It is far more useful for these purposes than as a medicine. Rosemary oil is widely employed as a fragrance component in soaps, creams, lotions, perfumes, and toilet waters; small amounts are also added as a flavoring agent to alcoholic beverages, frozen desserts, candy, puddings, and similar products. Extracts containing carnosic and labiatic acids have been shown to have antioxidant (food preservative) properties similar to those of butylated hydroxyanisole (BHA) and butylated hydroxytoluene (BHT). However, the larger quantities of the oil necessary for therapeutic purposes are not safe when taken internally and produce irritation of the stomach, intestines, and kidneys.[7] Besides, using rosemary as an abortifacient is certainly not a valid usage. Just what was old Mrs. Rabbit doing, selling that rosemary tea?

REFERENCES

1. M. Grieve. *A Modern Herbal,* Volume 2. Dover Publications, New York, 1971, pp. 681-683.

2. R. C. Wren and R. W. Wren. *Potter's New Cyclopaedia of Botanical Drugs and Preparations,* New Edition. Health Science Press, Hengiscote, England, 1975, p. 261.

3. M. Pahlow. *Das Grosse Buch der Heilpflanzen.* Gräfe und Unzer GmbH, Munich, Germany, 1979, pp. 270-272.

4. P. H. List and L. Hörhammer (Eds). *Hagers Handbuch der Pharmazeutischen Praxis,* Fourth Edition, Volume 6B. Springer-Verlag, Berlin, 1979, pp. 172-176.

5. A. Y. Leung and S. Foster. *Encyclopedia of Common Natural Ingredients Used in Food, Drugs, and Cosmetics,* Second Edition. John Wiley and Sons, New York, 1996, pp. 446-448.

6. *Bundesanzeiger,* November 30, 1985; March 6, 1990.

7. M. Pahlow, op. cit., pp. 270-272.

Royal Jelly

The newspaper headline read "Battling Budapest Baldies Briskly Buy Banfi," and the story that followed told how Hungarian men were engaging in brawls and fistfights to get a place in line so they might spend the equivalent of $2.85 to buy a bottle of the latest herbal hair restorer—a bottle that might sell on the black market for as much as $100. The exact formula of the "wonder" remedy concocted by promoter Andras Banfi was a secret, of course, but those in the know speculated that it contained egg yolk, orange tincture, alcohol, and royal jelly.[1]

Royal jelly is a milky white, viscous secretion produced by the pharyngeal glands of the worker bee, *Apis mellifera* L., an insect belonging to the family Apidae. During the first three days of life, all bee larvae feed on it exclusively. Future queens continue to be nourished by this interesting product that is somehow responsible for their development into mature females.[2]

Because the resulting queens are much larger than worker bees, live ten times longer, and are highly fertile (worker bees are sterile), enthusiasts have long hoped that royal jelly might have beneficial effects when consumed or applied externally by human beings. The product is commercially available in a wide variety of forms, including ampules, capsules, creams, lotions, soap, and the like.

Various writers have claimed that royal jelly is especially effective in halting or controlling the aging process—to nourish the skin and erase facial blemishes and wrinkles; in cases of fatigue, depression, convalescence from illness, the "growing pains" of adolescence; and in preventing the signs of normal aging or even premature senility. As a general tonic for treating the menopause or male climacteric and to improve sexual performance, royal jelly supposedly has a general systemic action rather than any specific biological function. The advertising brochure for a Chinese product also advocates its use in cases of liver disease, rheumatoid arthritis, anemia, phlebitis, gastric ulcer, degenerative conditions, and general mental or physical weakness.[4]

Although the chemistry of royal jelly is still not completely known, it has been extensively studied and found to contain protein, lipids, carbohydrates, fatty acids, and vitamins. The B vitamins were especially prominent, with pantothenic acid predominating. One component, sometimes called royal jelly acid (10-hydroxy-*trans*-2-decenoic acid) is considered a major component to which antibacterial activity has been attributed.[5] Tests have shown that royal jelly does possess some slight antibacterial activity; it can also affect the adrenal cortex and produce hyperglycemia (high blood sugar). An antitumor effect in mice has also been noted. However, there is no evidence that the product has any estrogenic (female sex hormonal) activity, or that it affects the growth, longevity, or fertility of animals.[6]

As for its topical effectiveness in rejuvenating the skin, the results from one three-month clinical study of twenty-four female patients are of interest. Ten women showed improvement, ten experienced no change, and four showed symptoms of skin irritation.[7] These are equivocal results at best.

In view of the lack of evidence to support these claims of therapeutic usefulness, we can only agree with Dayan.[8] Any value in human beings is purely psychological (attributable to the placebo effect) and springs from the novelty and glamour of treatment with such an exotic product. One physiological effect of royal jelly is indisputable. It does have the ability to produce queens from ordinary bee larvae. Prudent readers will limit its use to that purpose. Those baldies in Budapest are still flat on top and so are their wallets!

REFERENCES

1. *The Indianapolis Star,* March 15, 1979, p. 1.
2. V. E. Tyler, L. R. Brady, and J. E. Robbers. *Pharmacognosy,* Ninth Edition. Lea and Febiger, Philadelphia, Pennsylvania, 1988, p. 486.
3. S. S. Jones. *Herbalist* 4(10): 24-25, 1979.
4. *Peking Royal Jelly.* Peking Dietetic Preparation Manufactory, Peking, nd.
5. A. Y. Leung and S. Foster. *Encyclopedia of Common Natural Ingredients Used in Food, Drugs, and Cosmetics,* Second Edition. John Wiley and Sons, New York, 1996, pp. 448-451.
6. A. D. Dayan. *Journal of Pharmacy and Pharmacology* 12: 377-383, 1960.
7. J. S. Jellinek. *Formulation and Function of Cosmetics.* John Wiley and Sons, New York, 1970, pp. 393-394.
8. A. D. Dayan, op. cit., 377-383.

Rue

Fresh leaves of the small, yellow-flowered, evergreen shrub, *Ruta graveolens* L. (family Rutaceae) emit a strongly disagreeable odor, which, once smelled, will not be forgotten. Native to Europe, but naturalized and cultivated in the United States, this unpleasantly aromatic plant has been used since ancient times to prevent contagion (plague) and to repel insects as well as to heal their bites.[1] Dried rue leaves, which are less fragrant due to loss of much of their contained volatile oil, have also long been used as a folk remedy, particularly as an antispasmodic (to relieve cramps), a calmative, an emmenagogue (promotes the menstrual flow), and an abortifacient.

Rue does contain a number of active constituents. A mixture of quinoline alkaloids, present in the herb to the extent of 1.4 percent, and especially one designated arborinine, possess spasmolytic and abortifacient properties. Coumarin derivatives, a large number of which are present in the plant and in its volatile oil, also contribute appreciably to its spasmolytic properties.[2,3]

Unfortunately, these so-called furocoumarins, such as bergapten and xanthoxanthin, confer a significant toxicity especially on fresh rue, causing it to blister the skin following contact and exposure to sunlight (photosensitization).* An Indiana woman and her two young sons who rubbed fresh rue leaves on their exposed skin after reading about its insect repellent properties in a gardening magazine suffered from hives and large blisters (some three inches across) that required more than two weeks of treatment by a physician.[4] Foster has observed numerous cases of contact dermatitis from fresh rue, primarily suffered by gardeners who unwittingly handle the plant while weeding or transplanting. The fresh plant taken internally may result in gastric upsets, and Pahlow warns that these

---

*See the angelica monograph for additional comments on the toxicity of these so-called psoralens.

may also be caused by large doses of the dried leaves.[5] However, the toxicity of rue is much diminished on drying as a result of a decrease in the volatile oil content.

Several statements about rue in the current herbal literature need clarification. There is no evidence to support the assertion that any adverse symptoms from an overdose of the drug can be overcome by administering a small amount of goldenseal.[6] It is certainly doubtful that rubbing fresh rue leaves on the forehead will cure a headache,[7] but it is reasonably certain that, if exposed to the sun, this will result in a kind of dermatitis much worse than the original headache!

Although there is little question about the antispasmodic action of rue, there is appreciable doubt about the utility and safety of the drug, especially in the fresh state. German health authorities have concluded that neither rue nor any of its preparations should be utilized in medicine for two reasons:[8] In the first place, its utility is unproven. Second, there is a very unfavorable risk-benefit ratio for the herb. Belief in rue as a valuable medicinal agent is as ridiculous as the belief that if the gunflints for a flintlock muzzleloader were boiled in a mixture of rue and vervain, the shot would hit its mark no matter how poor the aim of the marksman.[9]

REFERENCES

1. H. N. Moldenke and A. L. Moldenke. *Plants of the Bible.* Chronica Botanica Co., Waltham, Massachusetts, 1952, p. 208.

2. P. H. List and L. Hörhammer (Eds.). *Hagers Handbuch der Pharmazeutischen Praxis,* Fourth Edition, Volume 6B. Springer-Verlag, Berlin, 1979, pp. 204-208.

3. A. Y. Leung and S. Foster. *Encyclopedia of Common Natural Ingredients Used in Food, Drugs, and Cosmetics,* Second Edition. John Wiley and Sons, New York, 1996, pp. 451-453.

4. J. Gengler. *Gary Post Tribune,* July 31, 1982, pp. Al, A3.

5. M. Pahlow. *Das grosse Buch der Heilpflanzen.* Gräfe und Unzer GmbH, Munich, Germany, 1979, pp. 266-267.

6. M. Tierra. *The Way of Herbs.* Unity Press, Santa Cruz, California, 1980, p. 113.

7. N. Coon. *Using Plants for Healing,* Second Edition. Rodale Press, Emmaus, Pennsylvania, 1979, p. 178.

8. *Bundesanzeiger,* January 5, 1989.

9. H. N. Moldenke and A. L. Moldenke, op cit., p. 208.

Sage

There is an old proverb:

> He that would live for aye,
> Must eat sage in May.

This aptly summarizes the folkloric belief in the leaves of common garden sage as an almost magical cure-all. This well-known plant with its many-branched stem, opposite hairy leaves, and blue, rarely pink or white, flowers is cultivated in the temperate parts of Europe and North America. Known by the scientific name *Salvia officinalis* L., it belongs to the mint family or Lamiaceae. Commercial samples are frequently adulterated with Greek sage, *Salvia triloba* L., another recent victim of taxonomic tinkering now referred to as *Salvia fruticosa* Mill.

Keller lists more than sixty different ailments for which sage is claimed to be therapeutic.[1] Alphabetically, these range from aches to wounds and include such conditions as congealed blood, falling sickness (epilepsy), insomnia, measles, rheumatism, seasickness, venereal disease, and worms. There is really little need to elaborate on these, for if one consults a wide enough variety of modern herbals, it is highly probable that every sickness known to man will be listed as being cured by sage—plus a special ability to strengthen the nerves, quicken the senses and the memory, and promote longevity.[2]

On a more rational level, sage is taken extensively as a household remedy in Europe for several purposes: an aid in drying up the flow of mother's milk at the end of the nursing period; internally to reduce the secretion of saliva; and particularly as an anhidrotic (reduces or stops perspiration) to control the night sweats associated with diseases such as tuberculosis. According to its supporters, various liquid preparations of sage relieve inflammations of the oral cavity and throat when used as a mouthwash or gargle.[3]

This last action on the mucous membranes is readily accounted for by sage's content of volatile oil and tannin. Its 1 to 2.8 percent of volatile oil, consisting mainly of α and β-thujones, has been shown to possess antiseptic properties; its condensed catechin-type tannin acts as an astringent and also stimulates blood flow by its local irritant properties. These actions combine to make sage useful in treating mouth and throat irritations. Unfortunately, some of the other medicinal properties of sage are more controversial and not so readily explained.

However, clinical studies carried out on sage tincture and sage tea as early as 1896 demonstrated the drug's ability to suppress perspiration. This action was confirmed by repeated experiments conducted during the early decades of this century which showed that perspiration was reduced by about one-half, maximum effect being achieved in about 2 to 2½ hours. The activity is attributed to constituents in the volatile oil. A proprietary preparation of it, Salvysat Bürger, is currently marketed in Germany as an anhidrotic.[4]

In 1939, experiments demonstrated estrogenic (female sex hormonal) activity following injection of sage extracts in mice. The compound(s) responsible for this effect remains unidentified. Furthermore, the relationship, if any, of this activity to the drug's purported ability to dry up the milk of nursing mothers is extremely unclear.

Other pharmacological investigations showed that administration of a decoction (boiled aqueous extract) of sage produced significant reduction of blood sugar in human subjects suffering from diabetes. The drug was particularly effective when taken on an empty stomach.[5]

All of these scientific studies would obviously confirm many of the uses of sage as a popular household remedy if it were not for an extensive investigation conducted in 1949 by H. B. J. van Rijn, which cast doubt, or at least raised a real question, concerning them. Using small animals, he was unable to demonstrate any anhidrotic effect in sage, nor could he detect effects on blood pressure and respiration. In addition, he could not show any antibacterial action of the drug, which also produced contradictory actions on various smooth muscles of the different organs (intestines and uterus). Finally, he did show a marked toxic effect of sage, probably attributable to its thujone-containing volatile oil.[6]

Details of the poisonous character of thujone are provided in the monograph on wormwood. Basically, it can cause both mental and physical deterioration when consumed in small amounts over a long period of time. Large doses can result in convulsions and loss of consciousness. We know that the plants containing it are not innocuous, so if van Rijn was right in this part of his study, he may also have been correct in his conclusion that sage, aside from its astringent action, is without medicinal value.

German health authorities have declared sage to be effective, when used as a mouthwash or gargle, for various inflammatory conditions of the mouth and throat.[7] This use appears justifiable. However, they also allow its internal use for digestive upsets and excessive perspiration. On the basis of published data, these latter uses seem inappropriate because of the known toxicity and lack of proven effectiveness of the herb in such cases. Steinegger and Hänsel note that the various medicinal preparations of sage contain relatively little volatile oil and only the purified oil is liable to cause toxic effects.[8] But after all, herbs should be used to improve health, not to diminish it. The cautionary statement of Wichtl, warning against the prolonged consumption of sage tea, seems quite appropriate.[9]

Even if it does have some therapeutic usefulness, sage cannot be recommended as a medicinal for internal use because of its high thujone content. Adding the leaves as a spice or flavoring agent in cooked foods is probably not critical since the heat involved apparently drives off most of the volatile thujone. Still, to be absolutely safe, we probably should reword that old adage a little, to read:

> He that would live for aye,
> Won't eat much sage in May—
> Or any other month.

REFERENCES

1. M. S. Keller. *Mysterious Herbs and Roots.* Peace Press, Culver City, California, 1978, pp. 300-314.

2. M. Israel. *Medical Herbalist* 9: 173, 1936-1937.

3. E. Steinegger and R. Hänsel. *Lehrbuch der Pharmakognosie,* Fourth Edition. Springer-Verlag, Berlin, 1988, pp. 343-345.

4. Rote Liste 1985, Editio Cantor, Aulendorf/Württ., 1985, index no. 31 270.

5. F. Berger. *Handbuch der Drogenkunde*, Volume 2. Verlag Wilhelm Maudrich, Vienna, 1950, pp. 292-305.

6. H. B. J. van Rijn. *Pharmaceutisch Weekblad* 84: 337-343, 1949.

7. *Bundesanzeiger*, May 15, 1985.

8. E. Steinegger and R. Hänsel. *Lehrbuch der Pharmakognosie,* Fourth Edition. Springer-Verlag, Berlin, 1988, pp. 343-345.

St. John's Wort

Because *Hypericum perforatum* L., an aromatic perennial herb belonging to the family Hypericaceae, produces golden yellow flowers that seem to be particularly abundant on June 24, the day traditionally celebrated as the birthday of John the Baptist, the plant is commonly known as St. John's wort. Its overground parts (leaves and flowering tops) that are medicinally applied also begin to be harvested at about that time.[1] The plant is native to Europe but is found throughout the United States.

St. John's wort was known to such ancient authorities on medicinal plants as Dioscorides and Hippocrates; indeed it is described and recommended as a useful remedy in all of the herbals down through the Middle Ages. But as with many plant drugs, it fell into disrepute in the late nineteenth century and was nearly forgotten. Quite recently, a tea prepared from the herb acquired a renewed reputation, particularly in Europe, as an effective nerve tonic, useful in cases of anxiety, depression, and unrest. Users also value it internally as a diuretic and in the treatment of various conditions, ranging from insomnia to gastritis.[2]

An olive oil extract of the fresh flowers of St. John's wort acquires a reddish color after standing in sunlight for several weeks. This so-called red oil is taken internally for the same conditions as is the tea, but it is also applied externally to relieve inflammation and promote healing. It is highly valued in the treatment of hemorrhoids.[3,4]

Chemical investigations have detected a number of constituents in St. John's wort, including about 1 percent of a volatile oil and approximately 10 percent of tannin. The latter compound probably exerts some wound-healing effects through its astringent and protein-precipitating actions. Much of the activity reported for the plant was initially thought to be due to the presence of hypericin, a reddish dianthrone pigment. Studies then tentatively linked the antidepressant effects of St. John's wort to various contained xanthones and flavonoids.[5] However, most recent investigations definitely suggest that other constituents in the whole extract, rather than hypericin and related compounds, are responsible for efficacy in mild to moderate forms of depression. Hyperforin is one currently being investigated.

The exact mechanism of action by which St. John's wort improves these depressive states is still unknown. It may involve the dopaminergic system. Other proposed mechanisms of action include an increase of neurotransmitters; inhibition of catechol-*0*-methyltransferase; modulation of cytokine activity; hormonal effects; and photodynamic effects.[6] It is quite possible that the herb functions by a variety of these, or similar, mechanisms, thereby explaining its minimal side effects.

In June of 1997, millions of American consumers self-diagnosed themselves as mildly to moderately depressed after a positive feature on the herb by ABC television news magazine program *20/20*. Product sales rose dramatically following the national media attention.

To date, at least twenty-seven controlled clinical studies have been published on St. John's wort, involving about 1,700 patients. Six studies were placebo-controlled. Ten studies have compared it with conventional drugs. Most studies have shown a positive benefit. Therapeutic use is approved in Germany for psychoautonomic disturbances, depressive mood disorders, anxiety and/or nervous unrest. At the time of publication of the German monograph in 1984, only one clinical study had been published. Since then, it has become clear that hypericin is no longer important in calibrating products. Therefore, the Federal Institute for Drugs and Medical Products in Germany no longer allows products to be labeled based on hypericin content.[7] Despite this fact, the vast majority of standardized products still available in the American marketplace are standardized to hypericin content, now considered a "marker" compound, usually at a level of 0.3 percent.

Unfortunately, hypericin may exert another action that is much less desirable. The compound is known to induce a kind of photosensitivity characterized by dermatitis of the skin and inflammation of the mucous membranes on exposure to direct sunlight.[8] While this is unlikely to happen with normal therapeutic doses of St.

John's wort, those who take large amounts of the herb for extended periods should be aware of the possibility and discontinue usage if such symptoms occur.

Depression can be serious, even life threatening. It is very important for consumers to receive proper diagnosis and to discuss thoroughly the use of St. John's wort with their health care provider.

REFERENCES

1. C. Hobbs. *HerbalGram* 18/19: 24-33, 1989.

2. P. Schmidsberger. *Knaurs Buch der Heilpflanzen.* Droemer Knaur, Munich, Germany, 1980, pp. 103-106.

3. M. Pahlow. *Heilpflanzen Heute.* Gräfe und Unzer GmbH, Munich, Germany, nd, pp. 65-66.

4. G. Mihailescu and A. Mihailescu. *Pflanzen Helfen Heil.* Biblio Verlagsgesellschaft mbH, Munich, Germany, 1979, pp. 56-57.

5. J. Hölzl. *Deutsche Apotheker Zeitung* 130: 367, 1990.

6. H. D. Reuter. In *Phytomedicines of Europe Chemistry and Biological Activity.* L. D. Lawson and R. Bauer (Eds.). American Chemical Society, Washington, DC, 1998, pp. 287-308.

7. V. Schulz, R. Hänsel, and V. E. Tyler. *Rational Phytotherapy: A Physicians' Guide to Herbal Medicine,* Third Edition. Springer, Berlin, 1998, pp. 50-65.

8. P. H. List and L. Hörhammer (Eds.). *Hagers Handbuch der Pharmazeutischen Praxis,* Fourth Edition, Volume 5. Springer-Verlag, Berlin, 1976, pp. 214-217.

Sarsaparilla

The roots of several woody climbing plants native to Central and South America constitute the drug sarsaparilla. All of them are species of the genus *Smilax*, belonging to the family Smilacaceae. Included are *S. aristolochiaefolia* Miller, known as Mexican sarsaparilla, *S. regelii* Killip and Morton, commonly referred to as Honduran sarsaparilla, *S. febrifuga* Kunth, or Ecuadorian sarsaparilla, as well as other undetermined species of *Smilax*.[1]

After it was introduced to Europe from the New World in the mid-sixteenth century, the drug was valued primarily as a treatment for syphilis.[2] This reputation, disguised under the terms "alterative" or "blood purifier," continued in medical circles well into the present century. As Mrs. Alice West of Jefferson, West Virginia, put it in an early-day patent medicine advertisement:

> I was all run down before I began to take Ayer's Sarsaparilla, but now I am gaining strength every day. I intend using the Sarsaparilla till my health is perfectly restored.[3]

Of course, this puts a slightly different light on the white-hatted cowboy hero who always strode to the bar in the Saturday afternoon B movie and, shunning the alcoholic beverages that were being drunk by the black-hatted villains, calmly said, "Give me a bottle of Sarsaparilla."

Sarsaparilla contains a mixture of saponins derived mainly from sarsapogenin and smilagenin.[4] The saponins have a strong diuretic action as well as some diaphoretic, expectorant, and laxative properties. In addition, the plant material is a moderately useful flavoring agent. Neither the whole drug nor its contained saponins is effective in the treatment of syphilis or as a "blood purifier."

In recent times, sarsaparilla has been widely promoted as one of the ingredients in various herbal combination products intended to serve athletes and bodybuilders as legal replacements for illegal steroidal drugs. To promote such usage, some distributors claim that sarsaparilla contains testosterone. As a matter of fact, that hormone has never been detected in any plant, including sarsaparilla. Advertisements also claim that the saponins in the herb are converted in some way in the body to allow them to function similar to anabolic steroids. This is also untrue.[5]

But the most deceptive practice of all with respect to sarsaparilla is the substitution for it, in some commercial herb products, of *Hemidesmus indicus* (L.) Schult.[6] Known as false sarsaparilla or as Indian sarsaparilla for its country of origin, this plant belongs to an entirely different plant family (Asclepiadaceae), and it does not contain the same saponins or other principal constituents found in sarsaparilla. Read the label carefully of any product said to contain sarsaparilla; the herb must be obtained from appropriate *Smilax* species originating in tropical America, not from *Hemidesmus.*

REFERENCES

1. V. E. Tyler, L. R. Brady, and J. E. Robbers. *Pharmacognosy,* Ninth Edition. Lea and Febiger, Philadelphia, Pennsylvania, 1988, p. 486.

2. F. A. Flückiger and D. Hanbury. *Pharmacographia.* Macmillan, London, 1879, pp. 703-712.

3. A. Hechtlinger. *The Great Patent Medicine Era.* Grosset and Dunlap, New York, 1970, p. 76.

4. V. E. Tyler, L. R. Brady, and J. E. Robbers, op. cit., p. 486.

5. V. E. Tyler. *Nutrition Forum* 5: 23, 1988.

6. M. Blumenthal. *Health Foods Business* 34(4): 58, 1988.

Sassafras

Fill me with sassafras, nurse,
And juniper juice!
Let me see if I'm still any use!

Donald Robert Perry Marquis
"Spring Ode"

Sassafras is a plant whose virtues are almost uniformly praised by modern herbalists. A tea prepared from the root bark of this native American tree, *Sassafras albidum* (Nutt.) Nees of the family Lauraceae, is widely recommended as a spring tonic and "blood thinner." The root bark was being used to treat fevers by the natives of Florida prior to 1512 and formed one of the earliest exports of the New World. It still enjoys a considerable reputation as a stimulant, antispasmodic, sudorific (sweat producer), depurative ("purifier") and as treatment for rheumatism, skin diseases, syphilis, typhus, dropsy (fluid accumulation), and so on.[1]

Much of the persistent reputation of sassafras may no doubt be attributed to its pleasant taste and aroma. It contains up to 9 percent of a volatile oil, which, in turn, consists of about 80 percent safrole. For ycars it was a valucd flavoring agcnt in root bccr and similar beverages. But as a result of research conducted in the early 1960s, safrole was recognized as a carcinogenic agent in rats and mice.[2] Sassafras bark, sassafras oil, and safrole are now prohibited by the FDA from use as flavors or food additives.

Unfortunately, sassafras continues to be collected, used, sold, and written about as an herbal remedy. No one really knows just how harmful it is to human beings, but it has been estimated that one cup of strong sassafras tea could contain as much as 200 mg of safrole, more than four times the minimal amount believed hazardous to humans if consumed on a regular basis.[3]

Some manufacturers, recognizing the attractive flavor and aroma of sassafras, have attempted to overcome its toxicity by preparing a safrole-free extract of the root bark. Such efforts were probably doomed to failure from the start since safrole is the major component responsible for the desirable odor and taste of the plant. However, an even more serious drawback has been revealed. Recent studies have shown that even safrole-free sassafras produced tumors in two-thirds of the animals treated with it.[4] Apparently, other constituents in addition to safrole are responsible for part of the root bark's carcinogenic activity.

As a matter of fact, a question was raised about the carcinogenicity of safrole in humans by a 1977 study carried out by toxicologists in Switzerland.[5] They were unable to demonstrate the formation of 1′-hydroxysafrole, the metabolite actually responsible for safrole's cancer-producing effect, when small amounts of safrole were given by mouth to human volunteers. On the other hand, this so-called proximate carcinogen was detected in the urine of rats when safrole was fed to those animals. The finding suggests that the toxicity of safrole in man and in small animals may differ.

However, the doses of safrole given to the human subjects in the Swiss study were extremely small (maximum 1.655 mg), and this may account for the failure of the human subjects to metabolize it to 1′-hydroxysafrole. More studies are definitely necessary before any final conclusion can be reached regarding the safety of sassafras as an herbal remedy.

Safrole has a number of industrial applications, including use as a fragrance in soaps and commercial cleansers. Some perfume components are also synthesized from safrole. A new use has emerged in the past decade. Clandestine producers of "designer drugs" have used safrole and isosafrole, derived from the oil of sassafras root bark, as a precursor in the manufacture of methylenedioxymethamphetamine (known as MDMA, ecstasy, XTC, and Adam). This illicit drug gained notoriety in the early 1990s, but it is not a new compound. The synthesis of MDMA from safrole was first described in a German patent published in 1912.[6]

An overriding consideration in this entire matter of the safety and efficacy of sassafras is that the plant material has no really significant medical or therapeutic utility. Sassafras oil, in common with a large number of volatile oils, does possess some mild counterirritant properties on external application, but beyond these, none of the claims of its supporters has been documented in the modern medical literature. Despite its pleasant flavor and its folkloric reputation as a useful tonic, prudent people will avoid this drug because of its potentially harmful qualities.

Modern scientific evidence compels us to revise Marquis' verse:

> Shun the sassafras, nurse,
> Bring juniper juice!
> It's one herb I can still misuse!

REFERENCES

1. J. U. Lloyd. *Origin and History of All the Pharmacopeial Vegetable Drugs, Chemicals and Preparations,* Volume 1. The Caxton Press, Cincinnati, Ohio, 1921, pp. 289-297.

2. *IARC Monographs on the Evaluation of the Carcinogenic Risk of Chemicals to Man* 1: 169-174, 1972.

3. A. B. Segelman, F. P. Segelman, J. Karliner, and D. Sofia. *Journal of the American Medical Association* 236: 477, 1976.

4. G. J. Kapadia, E. B. Chung, B. Ghosh, Y. N. Shukla, S. P. Basak, J. F. Morton, and S. N. Pradhan. *Journal of the National Cancer Institute* 60: 683-686, 1978.

5. M. S. Benedetti, A. Malnoë, and L. Broillet. *Toxicology* 7: 69-83, 1977.

6. L. G. French. *Journal of Chemical Education* 72: 484-491, 1995.

Savory

Ancient herbals commonly mention two savories: summer, which consists of the overground portions of *Satureja hortensis* L., and winter, obtained from *S. montana* L. These two aromatic members of the mint (Lamiaceae) family are small, widely cultivated garden plants with narrow leaves and pale lavender, pink, or white flowers. Summer savory, which is more highly prized as a spice and as a folk medicine, is an annual; winter savory is a perennial. For hundreds of years, both have enjoyed a reputation as sex drugs.[1] Summer savory was believed to increase desire (act as an aphrodisiac), and winter savory was believed to decrease the sex drive (anaphrodisiac). It is easy to see why summer savory became the more popular herb.

In modern folk medicine, summer savory is currently believed to benefit the entire digestive system. According to its believers, the herb acts as a carminative, an antiflatulent, an appetite stimulant, and also works in diarrhea. A tea prepared from the herb is considered beneficial as an expectorant and cough remedy.[2] One very interesting use of the tea in Europe is for excessive thirst in diabetics.[3] Many other therapeutic applications are listed by various herbalists, but most of these (for example, improving vision and curing deafness) are so farfetched that they are not worth repeating.[4]

An extremely valuable use of summer savory is as a spice. It gives an excellent flavor to beans and other legumes.[5] In fact, its German name is *Bohnenkraut* or bean herb. Both savories, as well as the aromatic volatile oils obtained from them, are much used in flavoring various kinds of sausages.

Summer savory contains from 0.3 to 2 percent of a volatile oil consisting of about 30 percent carvacrol, 20 to 30 percent *p*-cymene, and lesser amounts of numerous other constituents. The plant also contains 4 to 8.5 percent of tannin.[6] Other compounds have been identified in the herb, but none has any noticeable physiological activity.

Because of its content of carvacrol and *p*-cymene, the volatile oil confers a mild antiseptic property on summer savory. This apparently combines with the astringent effect of the contained tannin to make the plant of some little value in simple diarrhea. The oil is probably fairly effectual, especially when combined with hot water in the form of a tea, for minor throat irritations and mild digestive upsets. Besides, it tastes good and is relatively harmless, at least in moderate amounts. Summer savory is certainly a pleasant herb; just don't expect too much from it.

REFERENCES

1. M. S. Keller. *Mysterious Herbs and Roots.* Peace Press, Culver City, California, 1978, pp. 316-325.
2. M. Pahlow. *Das Grosse Buch der Heilpflanzen.* Gräfe und Unzer GmbH, Munich, Germany, 1979, pp. 93-97.
3. P. H. List and L. Hörhammer (Eds.). *Hagers Handbuch der Pharmazeutischen Praxis,* Fourth Edition, Volume 6B. Springer-Verlag, Berlin, 1979, pp. 295-299.
4. M. Grieve: *A Modern Herbal,* Volume 2. Dover Publications, New York, 1971, pp. 718-719.
5. M. S. Kelle, op. cit., pp. 316-325.
6. P. H. List and L. Hörhammer (Eds.), op. cit., pp. 295-299.

Saw Palmetto

If you bought a small paperback book for $2.00 and found a paragraph in it devoted to a berry that, taken regularly, would increase the size of female breasts, and also build sexual vigor, increase sperm production, reverse atrophy of the testes and mammary glands, and relieve catarrhal soreness of the genitourinary system, you might be tempted to try the berries—if, of course, you had these problems.[1] You might even drink, three times a day, a tea made from the saw palmetto berries and, as one person did after some time, write your syndicated pharmacy columnist to find out why it had not become necessary to buy a bigger bra.[2]

The berries responsible for this true-to-life scenario are the ripe fruits, fresh or more often partially dried, of *Serenoa repens* (Barb.) Small, also known as *S. serrulata* (Michx.) Nichols. Commonly called saw palmetto or sabal, the plant that produces these dusky red to brownish black berries is a fan palm (family Arecaceae), six to ten feet tall, with leaf clusters 2 to 2.5 feet across, each consisting of twenty or more leaf blades, and each blade ending in two sharp points. Growing in sandy soil from South Carolina to Florida and west to Texas, the plant forms great colonies in the wild. Commercial supplies of the berries historically come from the area around Cape Canaveral, Florida.[3] Today, wild harvest of the berries is conducted throughout much of central and southern Florida, with some commercial harvest in southern Georgia.

During the first half of this century, saw palmetto was frequently used in conventional medicine, mostly as a mild diuretic and as therapy for chronic cystitis; it was also considered good for enlargement of the prostate. The active constituent was supposed to be a volatile oil, but physicians began to question saw palmetto's efficacy, and in 1950, it was deleted from the listing of official drugs in *The National Formulary.*

Then, during the 1960s, investigators found relatively high concentrations of free and bound sitosterols in the dried berries. Various plant extracts as well as pure β-sitosterol, which was also isolated, exhibited estrogenic activity when *injected* into immature female mice. Although the activity was found to be relatively high compared to other estrogen-like compounds isolated from plants, it was rather low in comparison to the female sex hormones themselves. A saw palmetto extract was only about $^1/_{10,000}$ as potent as estradiol, and even pure β-sitosterol was less than $^1/_{10}$ as strong.[4]

More recent studies have concluded that saw palmetto exerts at least some of its effects not through its estrogenic properties per se but because it has an antiandrogenic (anti-male sex hormone) action. The compounds responsible for this property have not been identified, but they occur in the fraction obtained from the berries by extraction with nonpolar solvents, such as hexane. (That is, they are fat-soluble, not water-soluble, constituents.)[5]

Preparations of the herb are widely used in Europe today for the treatment of conditions associated with benign prostatic hyperplasia or BPH (nonmalignant enlargement of the prostate gland). The German health authorities have approved saw palmetto for this purpose, but preparations made from it must contain the lipophilic (fat-soluble) components of the drug.[6] Thus, little or no benefit for such a condition would be derived from a tea made from the herb since it would not contain a therapeutic dose of the water-insoluble active principles. Besides, in the United States, the FDA has banned the sale of all nonprescription "drugs" intended for the treatment of BPH because they have not received evidence proving them effective.[7] Nevertheless, saw palmetto preparations are widely sold as "dietary supplements."

At least seventeen clinical studies have been published on the use of saw palmetto extracts in the treatment of BPH. However, nine of them did not use control groups. Seven studies involving 490 patients did have a double-blind, placebo-controlled design.[8] A recent six-month, double-blind randomized equivalence study compared a saw palmetto extract (Permixon) with the conventional BPH drug finasteride (Proscar). The study enrolled 1,098 men over fifty years of age with moderate BPH. Both treatments were deemed of value in reducing symptoms of BPH in about two-thirds of patients. The saw palmetto extract produced significantly fewer complaints of decreased libido and impotence compared with the finasteride group.[9] The study has been criticized for lack of a placebo arm and absence of a placebo run-in period.[10]

Evidence for some of the other effects attributed to the herb is insubstantial. It apparently does have some beneficial anti-inflammatory properties, and the claim for breast enlargement could, just possibly, in certain cases derive from its antiandrogenic effect. However, other effects, such as sexual vigor (at least in the male) and increased sperm production, are exactly the opposite of what one might expect of an antiandrogen.

In 1998, advertisements began to appear promoting the use of saw palmetto to stimulate hair regrowth in men. This assertion was apparently based on the unwarranted assumption that because the prescription drug finasteride (Propecia) was useful for treating both BPH and hair loss, saw palmetto would also be effective for both conditions. There is absolutely no clinical evidence to support the use of the herb to prevent the loss of hair or to promote its regrowth.

Saw palmetto extracts are widely prescribed by urologists in France, Italy, and Germany for the supportive treatment of BPH. The condition is not amenable to self-diagnosis or unsupervised treatment. Consumers considering this therapeutic option in the form of dietary supplement products are strongly advised to discuss the matter with their physician.

REFERENCES

1. A. Gottlieb. *Sex Drugs and Aphrodisiacs.* High Times/Level Press, New York and San Francisco, 1974, p. 64.

2. J. Graedon. *The Indianapolis Star* April 15, 1981, p. 18.

3. H. W. Youngken. *Textbook of Pharmacognosy,* Sixth Edition. The Blakiston Co., Philadelphia, Pennsylvania, 1948, pp. 168-171.

4. M. I. Elghamry and R. Hänsel. *Experientia* 25: 828-829, 1969.

5. G. Harnischfeger and H. Stolze. *Zeitschrift für Phytotherapie* 10: 71-76, 1989.

6. *Bundesanzeiger,* January 5, 1989; February 1, 1990.

7. Anon. *American Pharmacy* NS30(6): 17, 1990.

8. V. Schulz, R. Hänsel, and V. E. Tyler. *Rational Phytotherapy: A Physicians' Guide to Herbal Medicine,* Third Edition. Springer, Berlin, 1998, pp. 226-228.

9. J.-C. Carraro, J.-P. Raynaud, G. Koch, G. D. Chisholm, F. D. Silverio, P. Teillac, F. C. Da Silva, J. Cauquil, D. K. Chopin, F. C. Hamdy, M. Hanus, D. Hauri A. Kalinteris, J. Marencak, A. Perier, and P. Perrin. *The Prostate* 29: 231-240, 1996.

10. D. Brown. *Quarterly Review of Natural Medicine* (Spring): 13-14, 1997.

Schisandra

One of the newer of the old drugs resurrected by the American herbal medicine industry is schisandra, or schizandra, the dried ripe fruit of *Schisandra chinensis* (Turcz.) Baill. (family Schisandraceae), a vine native to China. Its ancient folkloric use in China was as an antiseptic, astringent, tonic, and the like. During the last decade or so, Chinese doctors began using the drug to treat hepatitis, and a few studies have been done of its potential for liver-protective effects and the nature of its active constituents.

Western herbal advocates now acclaim schisandra as an "adaptogen," an agent supposedly capable of increasing the body's resistance to disease, stress, and other debilitating processes. Schisandra is said to "increase energy, replenish and nourish viscera, improve vision, boost muscular activity and affect the energy cells of the entire body." One marketer claims that its schisandra product can "help to combat damage that can lead to premature aging." Another notes that its product is "capable of providing a more healthy, active and longer lifespan." Schisandra ads also claim that it is effective against premenstrual syndrome, stimulates immune defenses, balances body function, normalizes body systems, boosts recovery after surgery, protects against radiation, counteracts the effects of sugar, optimizes energy in times of stress, increases stamina, protects against motion sickness, normalizes blood sugar and blood pressure, reduces high cholesterol, shields against infection, improves the health of the adrenals, energizes RNA-DNA molecules to rebuild cells, and "produces energy comparable to that of a young athlete."[1]

Limited studies of schisandra's effects have been carried out in small animals. An investigation conducted by L. Volicer and colleagues in Czechoslovakia in 1966 noted that the drug had a stimulating effect in low doses, but this was reversed with large doses. These actions are similar to those of nicotine.[2]

The constituents responsible for the liver-protective effects of schisandra are apparently lignans—molecules composed of two phenylpropanoid units. More than thirty of these have been isolated from schisandra, some twenty-two of which were tested in 1984 by the Japanese investigator H. Hikino for their ability to reduce the cytotoxic effects of carbon tetrachloride and galactosamine on cultured rat liver cells.[3] Most were found effective, and some were quite active. However, when galactosamine was used as a cytotoxic agent, the protective effects of the lignans were reduced at higher doses. Dr. Hikino concluded that the lignans of schisandra were themselves toxic to the liver when administered in large doses over a long period of time.

Subsequently, Japanese researchers have investigated the mechanism by which two of the lignans, wuweizisu C and gomisin A, exert their liver-protective effects. They found that both compounds functioned as antioxidants, thereby preventing the lipid peroxidation produced by harmful substances such as carbon tetrachloride. Since lipid peroxidation leads to the formation of harmful lesions in the liver, the two compounds did indeed exert a protective influence.[4] In the 1990s, several additional pharmacological studies have been published on the effects of schisandra or its isolated compounds on various parameters of liver function and/or protection. Human studies are absent in languages accessible to most American scientists.

However, the reported evidence to date on the stimulatory and liver-protective role of schisandra is somewhat equivocal and certainly preliminary in nature. To determine whether schisandra has practical value as a drug, long-term studies of safety and effectiveness at various dose levels—first in animals and ultimately in human beings—are definitely needed.[5]

REFERENCES

1. R. D. Marconi. *Let's Grow Younger.* Scientific Nutrition Press, Seal Beach, California, 1983 pp. 9-10.

2. L. Volicer, M. Sramka, I. Janku, R. Cape, R. Smetana, and V. Ditteová. *Archives Internationales de Pharmacodynamie et de Therapie* 163: 249-262, 1966.

3. H. Hikino. In *Natural Products and Drug Development,* P. Krogsgaard-Larsen, S. B. Christensen, and H. Kofod (Eds.). Munksgaard, Copenhagen, 1984, pp. 374-389.

4. Y. Kiso, M. Tohkin, H. Hikino, Y. Ikeya, and H. Taguchi. *Planta Medica* 51: 331-334, 1985.

5. *Lawrence Review of Natural Products,* July 1988.

Scullcap

That a nearly worthless and essentially inactive plant material could be recommended in a 1970 publication as a "useful tranquilizing herb"[1] and praised in an herbal revised in 1990 as "an excellent herb for almost any nervous system malfunction"[2] says much about the gullibility of human beings. Nevertheless, such is the case with scullcap (skullcap), the overground parts of the plant *Scutellaria lateriflora* L., a member of the family Lamiaceae. This plant is native to the United States, but several different species have been employed in medicine; *S. baicalensis* Georgi, a native of East Asia, is the one commonly utilized in Europe, the root of which is well-known in the traditional medicine system in Asia. All are rather similar, erect, perennial herbs that reach a height of about two feet.

Scullcap was introduced into American medicine in 1773 by Dr. Lawrence Van Derveer who used it to treat cases of hydrophobia. The name mad-dog herb stems from this. Subsequently, it came to be utilized primarily for its reputed tonic, tranquilizing, and antispasmodic effects. As such, it was a common ingredient in many proprietary remedies for "female weakness." The drug was officially in *The United States Pharmacopeia* from 1863 to 1916 and then in *The National Formulary* until it was dropped in 1947.[3]

In 1916, Pilcher tested an extract of scullcap for its effect on the contractility of the excised guinea pig uterus and found its slightly depressant properties to be the least active of the drugs tested. Since it had no effect in normal doses on the uterus of living animals, he concluded the drug lacked therapeutic value.[4] In 1957, Kurnakov studied the effects of extracts of two other species of scullcap, *S. galericulata* L. and *S. scordiifolia* Fisch., in various small animals. Neither had any effect on blood pressure in cats or rabbits, nor did they depress the central nervous system in frogs. They also failed to exert any antispasmodic activity.[5] The report by Usow, a year later, that a tincture of S. *baicalensis* produced a long-lasting decline in blood pressure in dogs is contradictory and requires verification.[6]

Various species of *Scutellaria* contain a number of flavonoid pigments, including baicalein, scutellarein, and wogonin, which might be thought responsible for the antispasmodic effects attributed to the whole drug. Tests in mice showed that baicalein had no detectable antispasmodic activity, while that of wogonin was very slight, about one-fifth that of papaverine hydrochloride.[7]

Over fifty years ago, at a time when scullcap was still officially recognized in *The National Formulary,* Wood and Osol aptly summarized its virtues or, rather, lack thereof:

> Scullcap is as destitute of medicinal properties as a plant may well be, not even being aromatic. When taken internally, it produces no very obvious effects, and probably is of no remedial value. . . .[8]

Lack of significant therapeutic activity is not the only problem associated with the use of scullcap. A report in the medical literature summarized the observed hepatotoxic (liver-damaging) effects of the herb in four women who had been consuming proprietary products supposedly containing it for the relief of stress.[9] This report raises another concern, namely, the identity of scullcap. Studies in Britain showed that many wholesalers there were substituting a species of *Teucrium* (germander) for scullcap.[10]

In the American herb trade, wild-harvested *Scutellaria* species are collectively known as "blue skullcap" and are often harvested without regard to species determination. A buyer interested in purchasing a lower-priced material may prefer "pink scullcap," represented by the pink-flowered, widespread adulterant *Teucrium canadense* L.

Common germander *Teucrium chamaedrys* L. has been established as a causative factor in several cases of acute hepatitis and a case of fatal hepatic dysfunction, resulting in a ban on its sale by the French Ministry of Health in April 1992.[11] This raises the question as to whether other species of *Teucrium* may cause liver damage.

Two cases of "skullcap" poisoning, including one fatality, were reported from the Riks Hospital in Oslo, Norway, in 1991.[12,13] Once again, it is unclear whether *Scutellaria lateriflora* was the offending botanical or whether adulteration with a *Teucrium* species was the culprit.

Deficiencies in activity, safety, and quality all make scullcap a good herb to avoid.

REFERENCES

1. M. J. Superweed. *Herbal Highs*. Stone Kingdom Syndicate, San Francisco, California, 1970, pp. 15-16.

2. H. Santillo. *Natural Healing with Herbs,* Eighth Printing. Hohm Press, Prescott, Arizona, 1990, pp. 174-175.

3. E. P. Claus. *Pharmacognosy,* Third Edition. Lea and Febiger, Philadelphia, Pennsylvania, 1956, pp. 219-220.

4. T. Sollmann. *A Manual of Pharmacology,* Seventh Edition. W. B. Saunders, Philadelphia, Pennsylvania, 1948, p. 406.

5. B. A. Kurnakov. *Farmakologiya i Toksikologiya* (Moscow) 206: 79-80, 1957.

6. P. H. List and L. Hörhammer (Eds.). *Hagers Handbuch der Pharmazeutischen Praxis,* Fourth Edition, Volume 6B. Springer-Verlag, Berlin, 1979, p. 340.

7. S. Shibata, M. Harada, and W. Budidarmo. *Yakugaku Zasshi* 80: 620-624, 1960.

8. H. C. Wood Jr. and A. Osol, *The Dispensatory of the United States of America,* Twenty-Third Edition. J. B. Lippincott, Philadelphia, Pennsylvania, 1943, pp. 965-966.

9. F. B. MacGregor, V. E. Abernathy, S. Dahabia, I. Cobden, and P. C. Hayes. *British Medical Journal* 299: 1156-1157, 1989.

10. J. D. Phillipson and L. A. Anderson. *Pharmaceutical Journal* 233: 80-82, 1984.

11. D. Larrey, T. Vial, A. Pauwels, A. Castot, M. Biour, M. David, and H. Michel. *Annals of Internal Medicine* 117: 129-132, 1992.

12. S. Leander and L. Skogstrøm. *Aflenposten* November 6, 1991.

13. R. J. Huxtable. *Annals of Internal Medicine* 117: 165-166, 1992.

Senega Snakeroot

Plant roots often assume a twisted, tortuous shape, so the name snakeroot is an apt one. Unfortunately, it is applied to so many different species (some of which are listed on page 13), that without a modifier the term is meaningless. Senega snakeroot, seneca snakeroot, or just plain senega refers to the yellow root of *Polygala senega* L., a perennial herb (family Polygalaceae) with small white flowers, native to the woodlands of eastern North America from southern Canada to South Carolina.[1] It is also commercially produced in Japan and India.[2]

It was one of the new remedies introduced into medicine after the discovery of America where the Seneca Indians valued it as a cure for rattlesnake bite. Although this usage was probably based purely on the "Doctrine of Signatures," senega subsequently enjoyed great popularity as a nauseant expectorant and was a common ingredient in syrups and similar preparations for coughs and colds. The drug's popularity subsequently declined, and in 1960, it was dropped from *The National Formulary.* Modern herbalists continue to praise its virtues as an expectorant, diaphoretic (promotes perspiration), sialagogue (increases the flow of saliva), and emetic. It is said to be particularly good for asthma and bronchitis.[3]

Fresh senega snakeroot has a pleasant odor reminiscent of wintergreen due to its content of approximately 0.1 percent methyl salicylate. The active ingredient, however, is a complex mixture of triterpenoid saponins in the root in a concentration ranging from 8 to 16 percent.[4] The saponins act by local irritation on the lining of the stomach, thus causing nausea, which in turn stimulates both bronchial secretions and the sweat glands. Large doses cause vomiting and purging.[5]

There is no question about the effectiveness of senega snakeroot as an expectorant. It is the subject of a German Commission E. monograph, describing its use for catarrh of the upper respiratory tract for its secretolytic and expectorant effects.[6] Senega snakeroot continues to be used in Europe as an ingredient in various syrups, lozenges, and tea mixtures for controlling coughs and related throat irritations. If it is utilized in any of these forms, one must be careful to follow the recommended dosage, or stomach upsets will follow. For this and other reasons, the drug is simply not included in any commercial preparations in the United States. *The Handbook of Nonprescription Drugs* lists a large number of cough syrups with their ingredients, none of which contains senega as an expectorant, so we can only conclude that safer and more effective cough treatments exist.[7] If you need an expectorant, it is probably easier and better to use something other than senega snakeroot.

REFERENCES

1. A. Osol and G. E. Farrar Jr. *The Dispensatory of the United States of America,* Twenty-Fourth Edition. J. B. Lippincott, Philadelphia, Pennsylvania, 1947, pp. 1018-1019.

2. N. G. Bisset and M. Wichtl (Eds.). *Herbal Drugs and Phytopharmaceuticals: A Handbook for Practice on a Scientific Basis.* Medipharm Scientific Publishers, Stuttgart, Germany, 1994, pp: 384-385.

3. L. Veninga and B. R. Zaricor. *Goldenseal, Etc.* Ruka Publications, Santa Cruz, California, 1976, pp. 167-169.

4. C. J. Briggs. *Canadian Pharmaceutical Journal* 121: 199-201, 1988.

5. A. Osol and G. E. Farrar Jr. *The Dispensatory of the United States of America,* Twenty-Fourth Edition. J. B. Lippincott, Philadelphia, Pennsylvania, 1947, pp. 1018-1019.

6. *Bundesanzeiger,* March 12, 1986.

7. K. J. Tietze. Cold, Cough, and Allergy Products. In *Handbook of Nonprescription Drugs,* Eleventh Edition. American Pharmaceutical Association, Washington, DC, 1996, pp. 133-156.

Senna

There was an Old Man of Vienna,
 who lived upon tincture of senna;
When that did not agree he took chamomile tea,
That nasty Old Man of Vienna.

Edward Lear
The Book of Nonsense

It is possible that the Old Man of Vienna who found senna so disagreeable was the forerunner of the modern housewife who had a similar experience with the drug. Tired of the customary caffeine beverages and seeking an alternative hot drink, she selected a package of senna, unlabeled as to its use, from the shelf of a health food store. This beverage proved to be a different-tasting one, so she drank several strong cups of it during breakfast the next morning. After spending that afternoon and evening in the bathroom and in bed suffering from diarrhea accompanied by nausea, intense griping, and subsequent dehydration, she called her physician. He told her what she had already learned from the sad experience. Senna is a potent cathartic and, in measured amounts, a common ingredient in many proprietary laxatives.

Is this minor tragedy exaggerated or far-fetched? Not at all! Several similar cases are cited in the literature.[1]

Senna or senna leaves are actually the dried leaflets of two species of *Cassia: C. senna* L., known as Alexandria senna, and *C. angustifolia* Vahl, known as Tinnevelly senna. Both plants are small shrubs of the family Caesalpiniaceae. The former species grows along the Nile in Egypt and Sudan, and the latter is extensively cultivated in southern and eastern India. Some authorities now consider both plants as belonging to a single species, *Senna alexandrina* Mill.;[2] however, there are distinct morphologic and histologic differences between the two species, so this revision has not been widely accepted by pharmacognosists. Senna was introduced into European medicine in the ninth or tenth century by the Arabs; its use as a native drug apparently antedates written records.[3]

The centuries-old practice of senna as a cathartic drug is based on its content of so-called dianthrone glycosides (1.5 to 3.0 percent), principally sennosides A and B with lesser amounts of sennosides C and D as well as other closely related compounds.[4] Although it is certainly possible to take senna in the form of a tea prepared from one to two teaspoons of leaves, it may prove difficult to adjust the dosage of such an unstandardized preparation. Consequently, many consumers will prefer to select one of the many over-the-counter syrups, tablets, or similar products containing standardized amounts of active principles along with appropriate directions for use. Such preparations often contain added aromatic materials that somewhat modify the undesirable nauseant and griping effects of senna.[5]

A number of best-selling laxative products that did contain phenolphthalein, which has recently been removed from the market because of a link to cancer in rodents, have been reformulated to contain senna as the active ingredient.[6]

In any case, remember that senna is a potent cathartic drug, not just a different-tasting tea. Habitual dosing with this or any other anthraquinone-containing laxative should be avoided, or excessive irritation of the colon may result.[7] Chronic abuse may cause electrolyte disturbances and fluid imbalances due to potassium loss and may interfere with or potentiate the activity of cardiac glycosides. It is important to carefully follow label instructions and warnings for appropriate formulations of this important drug.[8]

REFERENCES

1. Anon. *Morbidity and Mortality Weekly Report* 27: 248-249, 1978.

2. A. O. Tucker, J. A. Duke, and S. Foster. Botanical Nomenclature of Medicinal Plants. In *Herbs, Spices, and Medicinal Plants: Recent Advances in Botany, Horticulture, and Pharmacology*, Volume 4, L. E. Craker and J. E. Simon (Eds.). Oryx Press, Phoenix, Arizona, 1989, pp. 169-242.

3. V. E. Tyler, L. R. Brady, and J. E. Robbers. *Pharmacognosy,* Ninth Edition. Lea and Febiger, Philadelphia, Pennsylvania, 1988, pp. 64-66, 488-489.

4. A. Y. Leung and S. Foster. *Encyclopedia of Common Natural Ingredients Used in Food, Drugs, and Cosmetics,* Second Edition. John Wiley and Sons, New York, 1996, pp. 472-474.

5. M. Grieve. *A Modern Herbal,* Volume 2. Dover Publications, New York, 1971, pp. 734-737.

6. S. G. Stolberg. *The New York Times,* November 10, 1997.

7. M. Pahlow. *Das Grosse Buch der Heilpflanzen.* Gräfe und Unzer GmbH, Munich, Germany, 1979, pp. 426-427.

8. V. E. Tyler and S. Foster. In *Handbook of Nonprescription Drugs,* Eleventh Edition. American Pharmaceutical Association, Washington DC, 1996, pp. 697-698.

Spirulina

A favorite theme of futuristic fiction is humankind's struggle for survival on an overpopulated planet where people are reduced to eating mostly food prepared from cultivated algae in order to get adequate nourishment. But fact has a way of catching up with even the most imaginative fiction, so in the 1990s (George Orwell of *1984* fame must be smiling), we see advertisements for the blue-green alga spirulina—one hundred 500-mg tablets for about $15.00.

Spirulina Turpin is a genus of blue-green algae belonging to the family Oscillatoriaceae of the division Cyanophyta.[1] There are numerous species, and the commercially available products are ordinarily not identified as to source. However, Switzer notes that two species are currently utilized: *Spirulina maxima*, cultivated in Mexico, and *S. platensis,* cultivated in Thailand and California.[2] New spirulina production facilities have recently been established in Hawaii, India, and China. In 1996, the California operation of Earthrise Farms, a division of Dainippon Ink & Chemicals, produced nearly 500 metric tons of an estimated 2,400 metric tons produced worldwide.[3]

Spirulina has for years been used as a food by natives of Africa and Mexico. Collected from the bottoms of seasonally dried-up ponds and shallow waters in the north of Lake Chad, where it has long been eaten by the African natives, it is known as *dihe.*[4] Early inhabitants of Tenochtitlán, the present-day Mexico City, also collected a blue-green alga of unknown identity that they found palatable (presumably a *Spirulina* species) from nearby lakes. They called it *tecuitlatl.*[5] Much of the current commercial supply of spirulina comes from Lake Texcoco in Mexico, but as previously noted, the alga is also cultivated and harvested elsewhere.

Touted as the "super food of the future," spirulina is said to contain 50 to 70 percent of highly digestible protein with all of the essential amino acids. It is also reputed to have twice as much vitamin B_{12} as liver, in addition to many other B vitamins, vitamins A and E, and minerals. Low in fat, with a mild and palatable flavor similar to bean sprouts, spirulina has an intense green color that many consumers find objectionable.[6]

In addition to its food value, spirulina is called a "safe diet pill" and an exciting new way to lose weight safely and quickly. This is based on the alga's content of the amino acid phenylalanine, which, according to one theory, affects the appetite center of the brain. Another theory is that eating spirulina raises the blood sugar concentration enough to influence the same hunger center of the brain, causing it to suppress hunger pangs.[7] Spirulina enthusiasts also say it is an effective treatment for diabetes, hepatitis, cirrhosis of the liver, anemia, stress, pancreatitis, cataracts, glaucoma, ulcers, and loss of hair.[8]

Tests in rats have shown that a commercial spirulina preparation consisting of 50 to 70 percent protein, 4 to 7 percent moisture, 6.4 to 9 percent ash (minerals), and 13 to 16.5 percent carbohydrates was able to be the sole dietary source of protein for these animals.[9] The alga does contain protein with a satisfactory makeup of amino acids, although the sulfur-containing ones, methionine and cystine, tend to be limiting.[10] Depending on the species, dried spirulina does contain between 0.5 and 2 μg (microgram) per gram of vitamin B_{12}.[11] However, selective assay procedures suggest that more than 80 percent of the "vitamin B_{12}" in spirulina is, in fact, analogues of the vitamin which have no vitamin B_{12} activity in humans. This very low level of activity in spirulina compares to 0.2 to 1.8 μg of the vitamin contained in 1 g of liver.[12] The adult daily requirement of vitamin B_{12} is 5 to 6 μg. A novel sulfated polysaccharide has been isolated from *S. platensis* that, in its pure form, was found to inhibit replication in vitro of several enveloped viruses.[13]

Granting that spirulina may be an acceptable source of protein, the question then arises if it is an economical one. Assuming the present price of about $15.00 for one hundred 500-mg tablets, then spirulina costs 30¢ per gram. If its protein content is a maximal 70 percent, then 1 gram of spirulina protein costs about 43¢. Compare this with roast beef which, depending on the cut, may cost about $3.50 per pound, equivalent to just over ¾¢ per gram. Considering its protein content to average 33 percent, then 1 gram of beef protein costs about 2¼¢ or nearly one-twentieth the cost of spirulina protein! The alga is certainly not an economical food, and many will prefer the taste of roast beef, especially at one-sixth the price, except, of course, vegetarians.

Does spirulina have anything else to offer? The vitamins and minerals it contains are easily obtained from other, more economical food sources. What about its effectiveness as an appetite suppressant? Any digestible carbohydrate-containing food will cause an increase in blood sugar and a corresponding reduction in hunger.

Certainly there are more economical sources of carbohydrate than spirulina. There is no evidence to support the claim that phenylalanine is especially effective in reducing the appetite, and even if it were, that amino acid is readily available from a wide variety of more economical protein sources. In 1979, a Food and Drug Administration Advisory Panel reviewed spirulina and found no reliable scientific data to demonstrate that it is a safe and effective appetite suppressant.[14]

As for claims that spirulina is a kind of miracle cure for everything from diabetes to hair loss, examination of the evidence in supporting such assertions reveals that it is insubstantial at best. A lot more clinical evidence must be obtained before any such contention can be accepted—for example, the study on reduction of hair loss was carried out on one patient!

If the science-fiction writers are right, it may someday be necessary for human beings to eat blue-green algae in order to exist. Fortunately, that day has not yet arrived. With meat and other protein sources readily available at a mere fraction of the cost of spirulina, consuming an alga that offers no clear nutritional or therapeutic advantages seems neither useful nor rational.

REFERENCES

1. G. M. Smith. *The Fresh-Water Algae of the United States.* McGraw-Hill, New York, 1950, pp. 573-574.

2. L. Switzer. *Spirulina,* Third Edition. Proteus Corp., Berkeley, California, 1980.

3. M. Lerner. *Chemical Market Reporter,* December 16, 1996.

4. J. Leóard. *Nature* 209: 126-128, 1966.

5. W. V. Farrar. *Nature* 211: 341-342, 1966.

6. H. N. Cole. *Herbalist New Health* 6(4): 15, 19, 1981.

7. R. G. Smith. *National Enquirer,* June 2, 1981, p. 23.

8. C. Hills (Ed.). *The Secrets of Spirulina.* University of the Trees, Boulder Creek, California, 1980.

9. A. Contreras, D. C. Herbert, B. G. Grubbs, and I. L. Cameron. *Nutrition Reports International* 19: 749-763, 1979.
10. G. Clement. *Revue de l'Institut Pasteur de Lyon* 4: 103-114, 1971.
11. V. Herbert and G. Drivas. *Journal of the American Medical Association* 248: 3096-3097, 1982.
12. P. H. List and L. Hörhammer (Eds.). *Hagers Handbuch der Pharmazeutischen Praxis,* Fourth Edition, Volume 2. Springer-Verlag, Berlin, 1969, p. 690.
13. T. Hayashi and K. Hayashi. *Journal of Natural Products* 59: 83-87, 1996.
14. Anon. *Pharmacy Practice* 16: 82, 1981.

Suma

Things are seldom what they seem,
Skim milk masquerades as cream.

William Schwenck Gilbert
H. M. S. Pinafore, Act II

Suma, the dried root of a tropical plant native to the Amazon rain forests, is definitely a "skim-milk" herb. It purports to be derived from a species of the family Amaranthaceae, designated *Pfaffia paniculata* (Mart.) Kuntze. Brazilian natives are said to refer to suma as *para todo,* meaning "for all things." It was introduced into this country some years ago as Brazilian ginseng, presumably in an effort to trade on the reputation of a well-established herb to which it is not remotely related.[1]

Advocates promote suma as an immune enhancer or adaptogen, that is, an agent that helps the body adapt to stresses of all kinds by restoring or enhancing the natural immune system. Dr. Milton Brazzach at the University of Sao Paulo is reported to have tested suma on some 3,000 patients, many with serious diseases such as cancer and diabetes. He found it to have "great healing and preventative powers." However, detailed reports of these clinical trials have never been published.

Promotional literature indicates that the herb has been used in the Amazon for at least 300 years as a tonic, aphrodisiac (sexual stimulant), and remedy for cancer, diabetes, tumors, wounds, and skin problems.[2] If so, the drug has been a well-kept secret, for it does not appear in any of the numerous and extensive compilations of medicinal plants of the world, including G. Dragendorff's *Die Heilpflanzen* or *Hagers Handbuch der Pharmazeutischen Praxis,* Fourth Edition. Nor is it prominent as an adaptogen in the catalog of 590 species of Brazilian medicinal plants published by the Department of Health in that country in 1983.[3] Interestingly, ginseng is so listed there.

Japanese investigators have conducted chemical studies of suma root.[4,5,6] They succeeded in isolating and characterizing a new nortriterpene, designated pfaffic acid, and six new saponin derivatives of that acid, which were designated pfaffosides A, B, C, D, E, and F. Preliminary tests revealed that certain of the pfaffosides inhibited the growth of cultured tumor cell melanomas (B16). This preliminary indication of cytotoxicity is interesting but by no means indicative that the plant is a useful anticancer agent in humans. Additional studies are required to determine if these constituents possess a sufficiently selective degree of toxicity to render them safe and effective drugs. A preliminary pharmacological study by Italian researchers did show that an ethanolic extract of the root produced mild anti-inflammatory activity with analgesic effects, but did not reduce noninflammatory pain, raising questions about its possible mechanism of action.[7] Aside from this study, recent scientific literature is limited to a report that the powdered root produced occupational asthma after exposure during the process of manufacturing capsules.[8]

Often when unproven recommendations are made for an herbal remedy, it is possible to take some comfort in knowing that the plant has at least been used as a folk medicine for hundreds, if not thousands, of years without apparent adverse effects. It is not unreasonable to assume that such an herb is probably safe for consumption by normal persons. That is not the case with suma. Although we are told that it is an ancient remedy in Brazil, no confirmation of this statement appears in the standard medicinal plant literature. Testing of its safety (toxicity) has not taken place, and appropriate studies of its efficacy are also lacking. In view of these deficiencies, it is not possible to recommend utilization of this remedy for any condition.

REFERENCES

1. F. Murray. *Better Nutrition* 47(6): 17, 20, 1987.

2. A. L. McQuade. *Suma: New Hope from Old Mother Nature.* Premier Herbs, Santa Monica, California, nd, 2 pp.

3. G. González Ortega, E. P. Schenkel, M. L. Athayde, and L. A. Mentz. *Deutsche Apotheker Zeitung* 129: 1847-1848, 1989.

4. T. Takemoto et al. (six other authors). *Tetrahedron Letters* 24: 1057-1060, 1983.

5. S. Nakai et al. (six other authors). *Phytochemistry* 23: 1703-1705, 1984.
6. N. Nishimoto et al. (six other authors). *Tennen Yuki Kagobutsu Toronkai Koen Yoshishu* 30: 17-24, 1988.
7. G. Mazzanti and L. Braghioli. *Phytotherapy Research* 8: 413-416, 1994.
8. J. Subiza, J. L. Subiza, P. M. Escribano, M. Hinojosa, R. Garcia, M. Jerez, and E. Subiza. *Journal of Allergy and Clinical Immunology* 88: 731-736, 1991.

Tansy

Known to botanists as *Tanacetum vulgare* L., family Asteraceae, tansy has a long history of use in folk medicine. This strongly aromatic herb, which reaches a height of up to three feet and produces bright yellow flowers, is native to Europe but is naturalized and widely cultivated in the United States.

The dried leaves and flowering tops of tansy have been employed, usually in the form of a tea, as an anthelmintic (expels worms), tonic, stimulant, and emmenagogue (promotes menstrual flow—often a euphemism for promoting abortion). Tansy also makes a flavoring in cakes and puddings, especially those eaten at Easter. And it enjoys a considerable reputation as an insect repellent, especially for flies.[1]

Fresh tansy yields between 0.12 percent and 0.18 percent volatile oil that is extremely variable in its chemical composition, depending upon the specific source plants utilized. Indeed, scientists indicate that a number of chemical races of tansy exist which perpetuate their own distinctive composition of the oil, just as other plants breed true for flower color or a similar, more noticeable characteristic. It is generally agreed that the physiological actions attributed to the plant mainly come from the thujone content of the oil.[2] But some tansy oils are entirely free of thujone, and others contain as much as 95 percent of that compound.[3] For example, one strain has been found whose essential oil contains parthenolide, the sesquiterpene deemed responsible for antimigraine activity in feverfew (*Tanacetum parthenium*), while thujone was absent.[4] This composition is determined by the genetic makeup of the plant and is not appreciably influenced by environmental factors. Thus, the effect of any tansy preparation will be dependent on the chemical race represented since this determines the thujone content of the contained volatile oil. Without subjecting a specific plant sample to an analysis for thujone, it is impossible to estimate the proper dosage for a tansy preparation.

Moreover, as mentioned in the discussion on wormwood, thujone is a relatively toxic compound, capable of inducing both convulsions and psychotic effects in human beings. There are far more effective and much safer medicines than the thujone-containing tansy for expelling and destroying intestinal worms—the principal use of the plant in folk medicine.[5] In this enlightened era, there is absolutely no reason to utilize a potentially dangerous, toxic material of this sort as an emmenagogue-abortifacient. As a matter of fact, since more effective insect repellents are readily available, there is no real reason to use tansy for anything. Well, perhaps there is just one. Tansy is used as a flavoring agent in certain alcoholic beverages, including Chartreuse, but the resulting product must be thujone-free.

REFERENCES

1. M. Grieve. *A Modern Herbal*, Volume 2. Dover Publications, New York, 1971, pp. 789-790.

2. P. H. List and L. Hörhammer (Eds.). *Hagers Handbuch der Pharmazeutischen Praxis,* Fourth Edition, Volume 3. Springer-Verlag, Berlin, 1972, pp. 902-905.

3. E. Stahl and G. Schmitt. *Archiv der Pharmazie* 297: 385-391, 1964.

4. H. Hendriks and R. Bos. *Planta Medica* 56: 540, 1990.

5. M. Pahlow. *Das Grosse Buch der Heilpflanzen.* Gräfe und Unzer GmbH, Munich, Germany, 1979, pp. 264-265.

Tea Tree

When Captain James Cook dropped anchor off the coast of New South Wales in 1770, some of his sailors went ashore and prepared an aromatic tea from the leaves of a tree growing in the swampy lowlands there. For that reason, the tree became known as the tea tree, not to be confused with the common tea plant, which yields both black and green tea. Subsequently, it was learned that the leaves of the tea tree, *Melaleuca alternifolia* (Maiden and Betche) Cheel of the family Myrtaceae, had been long used by the aboriginals as a local antiseptic. Settlers in that area gradually began to use the leaves and the volatile oil obtained from them for the treatment of cuts, abrasions, burns, insect bites, athlete's foot, and similar conditions.[1] The oil has since become a very popular home remedy there, and its use is now spreading throughout the world.

Tea tree leaves contain about 2 percent of a pale lemon-tinted volatile oil with a strong but pleasant nutmeg odor. It is obtained from the leaves by steam distillation. About one-third of the oil is comprised of various terpene hydrocarbons (pinene, terpinene, cymene); the remainder consists largely of oxygenated terpenes, particularly terpinen-4-ol, which may constitute up to 60 percent of the total oil. Sesquiterpene hydrocarbons and oxygenated sesquiterpenes are also present.[2]

The oil possesses pronounced germicidal activity due primarily to the terpinen-4-ol. It is important that the commercial product not be derived from other *Melaleuca* species, some of which contain high concentrations of cineole, a skin irritant that also reduces the antiseptic effectiveness of terpinen-4-ol. A recent study found the oil effective in vitro against methicillin-resistant *Staphylococcus aureus* strains.[3] Although terpinen-4-ol is considered the primary active component, Australian researchers have confirmed that other components in the essential oil, including alpha-terpineol and linalool also contribute to its antimicrobial action.[4]

During World War II, 1 percent of tea tree oil was routinely incorporated in the machine "cutting" oils used in munitions factories in Australia. This is said to have greatly reduced the number of infections resulting from abrasions on the hands of workers caused by metal filings and turnings.[5] Modern clinical studies of the oil are neither numerous nor extensive, but some do indicate its possible value in treating various vaginal and skin infections.[6,7] A trial involving 124 patients has provided evidence of its effectiveness in the treatment of acne vulgaris.[8] A recent controlled trial of 117 patients found tea tree oil an effective topical treatment for toenail onychomycosis (caused by dermatophy, yeast or occasionally mold, infections).[9] The oil is also incorporated into shampoo used to rid pets of fleas.

Irritation may result in sensitive individuals from the local application of tea tree oil. Its topical use is not, however, generally associated with any toxicity. The oil has a future in the home and hospital environment in various antimicrobial applications, but it is not the miracle product that some of its promoters would have us believe.[10]

REFERENCES

1. P. Babny. *Health Foods Business* 35(7): 65-66, 1989.

2. P. H. List and L. Hörhammer (Eds.). *Hagers Handbuch der Pharmazeutischen Praxis,* Fourth Edition, Volume 5. Springer-Verlag, Berlin, 1976, p. 750.

3. C. F. Carson, B. D. Cookson, H. D. Farrelly, and T. V. Riley. *Journal of Antimicrobial Chemotherapy* 35: 421-424, 1995.

4. C. F. Carson and T. V. Riley. *Journal of Applied Bacteriology* 78: 264-269, 1995.

5. A. R. Penfold and F. R. Morrison. "Tea Tree" Oils. In *The Essential Oils,* Volume IV, E. Guenther (Ed.). D. Van Nostrand Company, Inc., New York, 1950, pp. 529-532.

6. *Lawrence Review of Natural Products,* January 1991.

7. A. L. Blackwell. *Lancet* 337: 300, 1991.

8. I. B. Bassett, D. L. Pannowitz, and R. St. C. Barnetson. *Medical Journal of Australia* 153: 455-458, 1990.

9. D. S. Buck, D. M. Nidorf, and J. G. Addino. *The Journal of Family Practice* 38: 601-605, 1994.

10. S. Foster. *The Herb Companion* 48-52, February/March, 1994.

L-Tryptophan

Oh sleep! it is a gentle thing,
Beloved from pole to pole!

Samuel Taylor Coleridge
The Rime of the Ancient Mariner, Pt. V

Sleep, although much sought after, is sometimes not easily achieved. A glass of warm milk taken at bedtime is probably one of the oldest sleep aids known, and its use by millions of people over hundreds of years testifies to its effectiveness. Thus, in 1965, when it was realized that the brain chemical serotonin played an important role in the regulation of sleep, it seemed possible that L-tryptophan, a precursor of serotonin, might function as a useful sedative.[1] Because casein, which makes up about 3 percent of cow's milk, consists of about 1.2 percent of L-tryptophan, the presence of that amino acid was thought to explain the beneficial effects of milk in treating insomnia.

However, it was soon realized that the approximately 100 mg of L-tryptophan found in a glass of cow's milk was only about one-tenth the amount needed to promote sleep in a healthy adult. Besides, the other amino acids present in milk casein compete with L-tryptophan for passage into the brain. Subsequent studies indicated that the sedative properties of milk must be due to other constituents, possibly a group of compounds known as beta-casomorphins, which are known to produce various opioidlike (narcoticlike) effects in the body. Research in this field is still in its initial stages but has already produced interesting results.[2,3]

Returning to L-tryptophan, it is one of the essential amino acids. That simply means it is one that the body itself cannot synthesize, and since it is essential for normal growth and development, our requirement of it must be supplied from external sources. Fortunately, it is present to the extent of 1 to 2 percent in many plant and animal proteins. The minimal daily requirement of L-tryptophan for normal adult human beings is 3 mg (± 30 percent) per kilogram of body weight. This is equivalent to approximately one-quarter gram daily for a 180-pound adult.[4]

Studies in the 1960s and 1970s showed that 1-gram doses of L-tryptophan reduced sleep latency (time taken to fall asleep) in both normal subjects and mild insomniacs. Doses larger than 1 gram did not produce any increased response. There was general agreement that consumption of the compound increased subjective "sleepiness," making one fall asleep faster and reducing waking time.[5] As a result of these findings, and in spite of the fact that the product was not approved for drug use, health food outlets, quick to recognize a potential article of commerce, began to market 100-mg to 667-mg tablets and capsules of the amino acid. L-tryptophan became very widely utilized as an over-the-counter sleep aid.

Then disaster struck. In the late summer of 1989, cases of a serious blood disorder known as the eosinophilia-mylagia syndrome (EMS) began to occur in otherwise healthy individuals who had consumed quantities of L-tryptophan primarily as a sleep aid. The FDA eventually ordered the recall of manufactured L-tryptophan in any form at all dosage levels.[6] However, it is estimated that it had already caused some 5,000 cases of EMS and twenty-seven deaths. Eventually, all of the suspect product was traced to a single manufacturer, Showa Denko K.K. in Japan. That organization had not only used a new genetically engineered bacterium to produce the amino acid but had also modified the customary steps used to purify the L-tryptophan.

Suspicion regarding the cause of EMS then shifted away from L-tryptophan itself—which, after all, is a commonly consumed amino acid—to a contaminant produced, or at least not removed, by the new production process. The structure of the responsible contaminant has now been identified with reasonable certainty. It is a dimer (double molecule) of tryptophan, specifically 1,1′-ethylidenebis (tryptophan).[7]

Consumption of L-tryptophan in relatively large amounts produces an increase in the chemical serotonin, for which it acts as a precursor, in certain cells of the brain, where it functions as a neurotransmitter.[8] This increase in serotonin levels not only tends to induce sleep but also may be useful in the treatment of certain kinds of pain, mental depression, and other behavioral conditions. As a matter of fact, when L-tryptophan was withdrawn from the market and became unavailable for drug use in Britain, some patients there who were being treated with it for chronic depression suffered relapses.[9]

This observation strengthens the need for additional research into the safety and efficacy of pure L-tryptophan as a potentially useful therapeutic agent. However, until the purity and safety of the marketed product can be assured, consumption of manufactured L-tryptophan in any form or any quantity must be avoided. In recent years, marketing hype that once swirled around this amino acid has quietly slipped into a deep slumber.

REFERENCES

1. E. Hartmann. *The Sleeping Pill.* Yale University Press, New Haven, 1978, pp. 162-181.

2. V. Brantl and H. Teschemacher. *Naunyn-Schmiedeberg's Archives of Pharmacology* 306: 301-304, 1979.

3. T. Taira, L. A. Hilakivi, J. Aalto, and I. Hilakivi. *Peptides* 11: 1-4, 1990.

4. C. Lentner (Ed.). *Geigy Scientific Tables,* Volume 1. Ciba-Geigy Limited, Basel, Switzerland, 1981, p. 235.

5. E. Hartmann, op. cit., pp. 162-181.

6. L. N. Gelb (Ed.). *FDA Drug Bulletin* 20(1): 2-3, 1990.

7. A. N. Mayeno et al. (six other authors). *Science* 250: 1707-1708, 1990.

8. C. R. Craig. In *Modern Pharmacology,* Third Edition, C. R. Craig and R. E. Stitzel (Eds.). Little, Brown and Company, Boston, 1990, pp. 391-392.

9. I. N. Ferrier, D. Eccleston, P. B. Moore, and K. A. Wood. *Lancet* 336: 380-381, 1990.

Uva Ursi

If learning is a process of repetition, no one should have difficulty remembering the name of the plant that yields this drug, *Arctostaphylos uva-ursi* (L.) Spreng. and its varieties *coactylis* and *adenotricha* Pern. and Macbr. *Arctostaphylos* means bearberry in Greek; *uva ursi* means bearberry in Latin, and the plant is often called bearberry in English. But just to throw in one confusing element, it is the dried leaves, not the berries, of the widely distributed, trailing evergreen shrub of the family Ericaceae that possess its medicinal properties.[1]

In folk medicine, uva ursi is a diuretic and astringent for diseases of the bladder and kidneys. It is supposed to impart tone to the urinary passages and also to exert an antiseptic action there. This is supposed to render the drug practicable in various inflammatory diseases of the urinary tract such as urethritis, cystitis, etc.[2]

Uva ursi contains about 5 to 12 percent of the phenolic glycoside arbutin, which hydrolyzes when taken to release hydroquinone, the principal antiseptic and astringent constituent of the plant.[3] Ursolic acid, a triterpene derivative, and isoquercitrin, a flavonoid pigment, also contribute to the diuretic action.[4] Bearberry contains large amounts (15 to 20 percent) of tannin, an undesirable constituent that tends to upset the stomach. Consequently, the leaves should not be extracted with hot water, as is normally the case in preparing a tea. Rather, it is better to pour cold water over them and allow them to stand twelve to twenty-four hours before drinking. This minimizes the tannin content of the beverage.

Arbutin, or more specifically, the hydroquinone derived from it, is a rather effective urinary antiseptic, but only if taken in large doses and if the urine is alkaline. This means that consumers should avoid eating acid-rich foods, including many fruits and their juices, sauerkraut, vitamin C, and similar products.[5] Consumers must also be aware that hydroquinone, in large doses, is toxic and may cause ringing in the ears, vomiting, convulsions, and collapse. However, since the recommended dose of uva ursi is 1 gram, three to six times daily (delivering an average of 400 to 800 mg arbutin daily), and doses as large as 20 grams have produced no adverse response in healthy individuals, there would seem to be minimal cause for concern.[6] Uva ursi is not recommended for children and pregnant or lactating women. If symptoms persist beyond two weeks, or worsen during treatment, medical advice is necessary.[7]

Uva ursi is an ingredient in practically all of the kidney- and bladder-type teas, large numbers of which are marketed in Europe. It appears to be a modestly effective urinary antiseptic and diuretic if properly employed. The wisdom of self-determining conditions in which it might prove helpful and then self-treating them is, of course, an individual matter.

REFERENCES

1. V. E. Tyler, L. R. Brady, and J. E. Robbers. *Pharmacognosy,* Ninth Edition. Lea and Febiger, Philadelphia, Pennsylvania, 1988, pp. 77, 491.

2. M. Grieve. *A Modern Herbal,* Volume 1. Dover Publications, New York, 1971, pp. 89-90.

3. A. Y. Leung and S. Foster. *Encyclopedia of Common Natural Ingredients Used in Food, Drugs, and Cosmetics,* Second Edition. John Wiley and Sons, New York, 1996, pp. 505-506.

4. D. G. Spoerke Jr. *Herbal Medications.* Woodbridge Press Publishing Co., Santa Barbara, California, 1980, pp. 30-31.

5. M. Pahlow. *Das Grosse Buch der Heilpflanzen.* Gräfe und Unzer GmbH, Munich, Germany, 1979, pp. 66-70.

6. T. Sollmann. *A Manual of Pharmacology,* Seventh Edition. W. B. Saunders, Philadelphia, Pennsylvania, 1948, p. 576.

7. ESCOP: Uvae Ursi Folium. In *Monographs on the Medicinal Uses of Plant Drugs.* Volume 5. European Scientific Cooperative on Phytotherapy, Exeter, United Kingdom, 1997.

Valerian

Valerian and its extracts are contained singly and in combination in literally scores of drugs and teas that are currently available on the European market. Consisting of the dried rhizome and roots (underground parts) of *Valeriana officinalis* L. of the family Valerianaceae, the drug continues to be used after more than 1,000 years as a valued tranquilizer and calmative in cases of nervousness and hysteria. Other species of *Valeriana* also contain active principles and are similarly utilized.[1]

The valerian or garden heliotrope is a tall perennial herb whose hollow stem bears opposite leaves and white or reddish flowers. It has a vertical rhizome with numerous attached roots, which are harvested in the autumn of the second year's growth. These parts possess an extremely characteristic, disagreeable aroma arising from the contained volatile oil. The odor is said to be attractive to rats. (Legend has it that the Pied Piper used valerian to lure these pesky rodents from the village of Hamlin.[2])

From the 1960s through the 1980s, extensive studies were carried out on the nature of the active principles of valerian. By the mid-1980s, authorities were in general agreement that the herb's sedative effects were due to a mixture of unstable iridoid compounds known as valepotriates and to the contained volatile oil.[3] There was, however, considerable evidence against such a conclusion. The valepotriates are highly unstable, and most valerian preparations contained those originally present in the plant only in small amounts. Further, they are not well absorbed following oral administration. Water extracts containing no valepotriates and little volatile oil also proved to be effective sleep aids.[4]

Then, in 1988, Krieglstein and Grusla published the results of an extensive study in rats that proved what scientists had begun to suspect.[5] Although valerian was effective in producing depression of the central nervous system, neither the tested valepotriates, nor the sesquiterpenes valerenic acid, or valeranone, nor the volatile oil itself displayed any activity. At the present time, the identity of the principles responsible for valerian's therapeutic effects remains unknown. Possibly a combination of volatile oil components, valepotriates or their derivatives, and as-yet unidentified water-soluble constituents is responsible. There appears to be little doubt that the fresh root or that which has been recently and carefully dried at low temperature (under 40° C) constitutes the highest quality herb.

Because of their epoxide structure, some of the valepotriates demonstrate alkylating activity in vitro, that is, in cultured cells, and for a time, some concern was displayed for this potential toxicity. However, because the compounds decompose rapidly in the stored drug and are not readily absorbed, there is little cause for anxiety. Such toxicity has not been demonstrated in vivo, that is, in intact animals or human beings.

German health authorities have indicated that valerian is an effective treatment for restlessness and for sleep disturbances resulting from nervous conditions.[6] Ten controlled clinical studies have been conducted on various valerian preparations, including freeze-dried aqueous extracts, ethanol extracts, and other forms. Two studies involved only one test dose prior to evaluation. A recent German study showed no effects over the short term, but after twenty-eight days of treatment, valerian was shown to be better than a placebo in inducing sleep. This suggests that valerian root preparations may not be suitable for acute insomnia. More studies must be conducted to resolve differences in results with previous research.[7] Valerian is perhaps best characterized as a minor tranquilizer. The herb may be administered in the form of a tea, a tincture (hydroalcoholic solution), an extract, capsules, or tablets. It is also added to bath water for external application. No significant side effects or contraindications are noted, although rare reports of gastrointestinal complaints or headache have been reported.

Because of the similarity in spelling of the names and because both have a tranquilizing action, valerian and Valium are sometimes confused by consumers. The two products are, of course, very different. Valium is the trade name of a potent synthetic drug (it occurs in nature only in trace amounts), known generically as diazepam.[8] Valerian is a much milder tranquilizer of plant origin.

REFERENCES

1. H. W. Youngken. *Textbook of Pharmacognosy,* Sixth Edition. The Blakiston Co., Philadelphia, Pennsylvania, 1948, pp. 852-856.

2. W. H. Hylton (Ed.). *The Rodale Herb Book.* Rodale Press Book Div., Emmaus, Pennsylvania, 1974, pp. 611-613.

3. V. E. Tyler. *The New Honest Herbal.* George F. Stickley Company, Philadelphia, Pennsylvania, 1987, p. 231.

4. C. Hobbs. *HerbalGram* 21: 19-34, 1989.

5. J. Krieglstein and D. Grusla. *Deutsche Apotheker Zeitung* 128: 2041-2046, 1988.

6. *Bundesanzeiger,* May 15, 1985; March 6, 1990.

7. V. Schulz, R. Hänsel, and V. E. Tyler. *Rational Phytotherapy: A Physicians' Guide to Herbal Medicine,* Third Edition, Springer, Berlin, 1998, pp. 73-81.

8. U. Klotz. *Lancet* 335: 922, 1990.

Wild Yam

The genus *Dioscorea* of the family Dioscoreaceae includes more than 850 species of annual, twining, tuberous vines found in tropical and warmer temperate climates. Collectively, those species with edible roots are known as "yams." Some species with starchy fleshy tubers have been eaten as food. Others have been collected for their glycosides, particularly botogenin and diosgenin, whose steroidal nucleus is altered in the laboratory to produce numerous steroid hormones. The richest source of steroidal precursors is *D. floribunda* M. Martens and Galeotti, native to Mexico. *D. composita* Hemsl., also from Mexico, has served as a primary source of Mexican yam.[1,2] The so-called "wild yam," *D. villosa* L., best known historically as "colic root" is an herbaceous vine of rich open woods and moist thickets from Connecticut and New York, south to Florida, west to Texas and Minnesota. *D. quaternata* (Walter) J. F. Gmelin has a similar range and is separated on technical details. It is unlikely that they are distinguished by wild collectors of the roots.

Diosgenin was first isolated from *Dioscorea* by Japanese researchers in 1936. Conversion of this steroidal sapogenin to progesterone first occured in 1940. In the decades that followed, diosgenin and botogenin served as the major commercial precursors for the manufacture of steroid drugs, including oral contraceptives, topical homones, systemic corticosteroids, androgens, estrogens, progestogens, and other sex hormones.[3] Indirectly, *Dioscorea* species have quietly had a greater impact on twentieth-century social and medical practices than any other plant group.

Historically, the American wild yam *D. villosa* was little used by the medical profession. A botanic physican, Dr. Bone, one of the Hessian mercenaries captured by George Washington in the battle of Trenton in December of 1776, who later settled in New Jersey, is believed to have "discovered" that *D. villosa* was a remedy for "bilious colic," hence the name colic root. Previously, a poultice of the root had been used to treat contusions.[4]

In recent years, numerous "wild yam" and "Mexican wild yam" products, including both oral and topical dosage forms have appeared in the market. At least one multilevel company has promoted wild yam as a "natural precursor" to the hormone DHEA, claiming that diosgenin is converted by the body into this hormone. Of course, no scientific evidence is available to support this wishful thinking. Other promoters claim that application of a wild yam cream to the skin magically transforms disogenin into progesterone in the body. Some products add "natural" progesterone in an attempt to "deliver the goods." All of this is offered to reverse aging, treat menopausal and menstrual symptoms, and cure osteoporosis, to name just a few purported uses! According to one promotional brochure, "the Mexican yam," *Dioscorea villosa,* was first mentioned by the Chinese in 25 B.C.![5] How interesting that the Chinese supposedly knew, at this early date, about a so-called Mexican plant that does not even occur in Mexico.

Although diosgenin can be converted in the laboratory to numerous steroidal compounds, that chemical synthesis cannot and does not occur in the human body. The 1990s' wild yam scam is a sign of the urgent need for better self-regulation by the herb industry, to say nothing of more active federal regulators.

REFERENCES

1. D. J. Mabberley. *The Plant Book,* Second Edition. Cambridge University Press, New York, 1997, p. 321.

2. V. E. Tyler, L. R. Brady, and J. E. Robbers. *Pharmacognosy,* Ninth Edition. Lea and Febiger, Philadelphia, Pennsylvania, 1988, pp. 70-71.

3. J. M. Morton. *Major Medicinal Plants—Botany, Culture and Uses.* Charles C. Thomas Publisher, Springfield, Illinois, 1977, pp. 75-82.

4. D. E. Smith. *Transactions of the Eclectic Medical Society of New York,* 1870, pp. 623-628.

5. K. Keville. *The American Herb Association Newsletter* 12: 7, 1996.

Witch Hazel

Although only a small tree, *Hamamelis virginiana* L. (family Hamamelidaceae) is a very noticeable one, particularly in the fall. At a time when it and other trees begin to lose their leaves, witch hazel is suddenly covered with a multitude of golden yellow threadlike flowers. These often remain after the other autumn colors have disappeared, making the tree very conspicuous. Another North American species, vernal witch hazel, *H. vernalis* Sarg., indigenous to the Ozark plateau, also enters the herb trade as "witch hazel" without regard to species designation.

Witch hazel is a native American plant, and the topical use of its leaves or bark as a poultice to reduce inflammation was apparently introduced to early settlers by the Indians.[1] Various extracts were later employed both internally and externally for their astringent properties in conditions ranging from diarrhea to hemorrhoids. Then, about the middle of the nineteenth century, a very different kind of witch hazel preparation was introduced. Prepared by steam distilling the dormant twigs of the plant and adding alcohol to the aromatic distillate, it was designated hamamelis water, distilled witch-hazel extract, or just plain "witch hazel." The product was intended for local application to various skin conditions; large quantities are still marketed.

Tannin is the principal active ingredient in witch hazel; the leaves contain 8 percent, the bark from 1 to 3 percent. The tannin has been shown to consist of hamamelitannin and a number of proanthocyanidins. An important factor in the quaity of witch hazel extracts is the plant part used. The bark was found to be thirty-one times richer in hamamelitannin than the leaf extract.[2] A number of other constituents, including various flavonoid pigments are also present, but whatever astringent action the drug possesses seems to be accounted for by the tannin.[3]

Recently hamamelitannin and proanthocyanidins isolated from witch hazel were evaluated for their mechanisms of action in reported anti-inflammatory activity. It was found that some proanthocyanidin fractions inhibit inflammatory mediators derived from arachidonic acid and inhibit the formation of platelet-activation factor, also involved in the inflammatory process.[4] Strong antioxidant activity against superoxides, released by several enzymes during the inflammatory process, may also play a role in witch hazel's anti-inflammatory effects.[5] A recent controlled clinical study, which compared a witch hazel distillate (tannin-free) cream against 0.5 percent hydrocortisone cream in severe atopic eczema, found that the witch hazel cream was no better than its base preparation in reducing various symptoms of atopic eczema.[6]

In Europe, an alcoholic fluidextract of witch hazel is often taken internally to treat varicose veins.[7] Experiments on rabbits have shown that the drug does cause constriction of the veins, at least following injection.[8] The constituent(s) responsible for this activity remains unidentified. Interestingly enough, an alcoholic extract of the leaves was found to be much more active than an aqueous extract. Thus, anybody drinking tea prepared from the witch hazel bark commonly sold in health food stores should not expect much venous-constricting effect from it.

Extracts of witch hazel leaves and bark are the subject of a positive German Commission E monograph recognized for antiphlogistic, hemostyptic, and astringent properties for the topical treatment of skin injuries, burns, varicose veins, and hemorrhoids.[9] It should be noted that alcoholic extracts of the leaves and bark used in various European phytomedicine products are a completely different delivery form than the tannin-free witch hazel distillate prevalent in the American market.

Aside from its astringency and mild anti-inflammatory activity, it is a mistake to expect much of anything in the way of useful therapeutic action from this plant. At a time when various hamamelis preparations were still listed in *The National Formulary,* one authority commented, "Hamamelis is so nearly destitute of medicinal virtues that it scarcely deserves official recognition."[10] Hamamelis leaf and the fluidextract prepared from it were dropped from the 1955 edition of the NF.

Hamamelis water is especially interesting in that, due to its method of preparation by distillation, the final product is devoid of tannin; it is therefore essentially a mixture of 14 percent alcohol in water with a trace of volatile oil. The same authority just cited stated that hamamelis water fulfills "the universally recognized need in American families for an embrocation (liniment) which appeals to the psychic influence of faith."[11] Any astringent action exerted by the preparation is due to its alcohol content, which approximates that of table wine. Despite its doubtful utility, the FDA has declared that "witch hazel" (the distilled preparation) is a safe and effective astringent drug. Although red wine is seldom applied externally, it at least contains some tannin, and its therapeutic value as an astringent would therefore exceed that of "witch hazel."

REFERENCES

1. J. U. Lloyd. *Origin and History of all the Pharmacopeial Vegetable Drugs, Chemicals and Preparations,* Volume 1. The Caxton Press, Cincinnati, Ohio, 1921, p. 162.
2. B. Vennat, H. Pourrat, M. P. Pouget, D. Gross, and A. Pourrat. *Planta Medica* 54: 454-457, 1988.
3. P. H. List and L. Hörhammer (Eds.). *Hagers Handbuch der Pharmazeutischen Praxis*, Fourth Edition, Volume 5. Springer-Verlag, Berlin, 1976, pp. 9-14.
4. C. Hartisch, H. Kolodziej, and F. Bruchhausen. *Planta Medica* 63: 106-110, 1997.
5. C. A. J. Edelmeier, J. Cinatl Jr., H. Rabenau, H. W. Doerr, A. Bilber, and E. Koch. *Planta Medica* 62: 241-245, 1996.
6. H. C. Korting, M. Schafer-Korting, W. Klovekorn, G. Klovekorn, C. Marin, and P. Laux. *European Journal of Clinical Pharmacology* 48: 461-465, 1995.
7. P. Schauenberg and F. Paris. *Guide to Medicinal Plants.* Lutterworth Press, Guildford, England, 1977, pp. 291-292.
8. P. Bernard, P. Balansard, G. Balansard, and A. Bovis. *Journal de Pharmacie de Belgique* 27: 505-512, 1972.
9. *Bundesanzeiger,* August 13, 1985; corrected March 3, 1990.
10. A. Osol and G. E. Farrar Jr. (Eds.). *The Dispensatory of the United States of America,* Twenty-Fourth Edition. J. B.. Lippincott, Philadelphia, Pennsylvania, 1947, pp. 528-530.
11. Ibid.

Wormwood

> As he watered the green stuff in his glass,
> and the drops fell one by one.
>
> Robert Service
> "The Shooting of Dan McGrew"

Long before the "man from the creeks" had filled Dangerous Dan McGrew full of lead while under its influence, the herb known as wormwood or absinthe had acquired a sinister reputation. Although native to Europe, this shrubby, odorous plant, *Artemisia absinthium* L. of the family Asteraceae, has been naturalized in the United States and occurs widely in the northeastern and north central regions.[1]

As its common name wormwood implies, the herb was once used as an anthelmintic to destroy intestinal worms. Its leaves and flowering tops were also used as an aromatic bitter or tonic, a diaphoretic, and a flavoring agent. The late Euell Gibbons recommended three different formulas for wormwood preparations, including one which would cause the user to dream of his true love.[2] For reasons I shall explain, none of these can be endorsed. Wormwood is still employed in small amounts to flavor some of the aromatic alcoholic beverages, including vermouth, and to impart a fragrance to certain liniments.

Wormwood acquired its sinister reputation as a subtle poison when it became the principal flavoring ingredient in a 136-proof alcoholic beverage called absinthe. This green-colored aperitif was too strong to drink straight, so most tipplers diluted it with water as described in Service's poem. It was the "in" beverage served at all the sidewalk cafés in Paris around the turn of the century. Then, one absinthe addict, Vincent van Gogh, sliced off his own ear and mailed it to a lady friend; another, John Lanfray, murdered his pregnant wife, two daughters, and then attempted suicide while under its influence; and the impressionist, Edgar Degas, painted a truly haunting portrait of two hollow-eyed absinthe drinkers, seated at a table, oblivious to all around them as a result of the toxic beverage. Practically every civilized country in the world banned the preparation or consumption of absinthe. France, which prepared most of it and which consumed two-thirds of the world's supply, was among the last to do so, in 1915.

The principles absinthin and anabsinthin are responsible for the bitter taste of wormwood, but its pleasant aroma is due to a volatile oil that is contained in the herb in a concentration ranging between 0.25 and 1.32 percent. The oil, in turn, contains 3 to 12 percent of thujone (a mixture of the α- and β-forms), long believed to be the major toxic constituent in the plant.[3]

More than fifty years ago, the injection of thujone at levels as low as 40 mg per kg induced convulsions in rats, and caused fatalities when this quantity was increased to 120 mg per kg.[4] One-half ounce of wormwood volatile oil also caused convulsions and unconsciousness in a human being, according to an ancient medical report.

Only in the past twenty-five years have scientists been able to offer an explanation for the difference between the relatively large doses of thujone required to produce toxic effects in rats and the much smaller amounts known to impair the faculties of human beings. They propose that thujone exerts its psychotomimetic (mind-altering) effects by reacting with the same receptor sites in the brain as those which interact with THC (tetrahydrocannabinol), the active principle of marijuana.[5]

This hypothesis is supported by observing the same mind-altering effects induced by drinking absinthe and by smoking marijuana. Also, thujone and THC not only have similar molecular geometries and similar functional groups allowing them to "fit" a common receptor site without changing orientations or relative positions; they also are capable of similar types of oxidative reactions. The theory requires experimental verification, but it does explain why absinthe, even when consumed in relatively small amounts, could cause such profound mental and physical changes in habitual or even casual users.

Wormwood has no place in modern phytomedicine.

REFERENCES

1. H. W. Youngken. *Textbook of Pharmacognosy,* Sixth Edition. The Blakiston Co., Philadelphia, Pennsylvania, 1948, pp. 873-874.

2. E. Gibbons. *Stalking the Healthful Herbs,* Field Guide Edition. David McKay Co., New York, 1966, pp. 42-46.

3. H. A. Hoppe. *Drogenkunde,* Eighth Edition, Volume 1. Walter de Gruyter, Berlin, 1975, pp. 119-120.

4. W. L. Sampson and L. Fernandez. *Journal of Pharmacology and Experimental Therapeutics* 65: 275-280, 1939.

5. J. del Castillo, M. Anderson, and G. M. Rubottom. *Nature* 253: 365-366, 1975.

Yellow Dock

For hundreds of years, herbalists have been recommending the root of various species of dock for diseases of the blood and liver. These recommendations are repeated, using slightly different terminology, in modern herbal writings that describe yellow dock, *Rumex crispus* L. (family Polygonaceae), as a helpful alterative and laxative.[1] The term alterative refers to a drug intended for the treatment of syphilis and related venereal diseases; it is often used synonymously with "blood purifier."

Yellow dock is a perennial herb, growing up to about four feet in height, with slender leaves characterized by wavy-curled margins. This accounts for another widely used name for the plant, curly dock. It is a native of Europe but is found growing abundantly in waste places throughout most of the United States. The deep yellow, underground parts (rhizome and roots) make up the drug, but dock greens are also eaten as a potherb. Actually, a number of closely related species are similarly employed, and when Rumex was listed in *The National Formulary*, *R. obtusifolius* L. was also designated as a source.[2]

A number of anthraquinone derivatives, including chrysophanic acid, emodin, and physcion, among others, have been identified in yellow dock.[3] These account for the drug's laxative action, which is well substantiated. In fact, one study showed that the total anthraquinone content of this plant's root, 2.17 percent, exceeded the 1.42 percent concentration of these principles in medicinal rhubarb (not to be confused with garden rhubarb, which contains only small amounts of anthraquinones).[4] Incidentally, rhubarb belongs to the same plant family as yellow dock; many members of the Polygonaceae contain anthraquinones accompanied by significant amounts of tannin.

It is difficult to understand how a simple laxative drug could have retained its ancient reputation for being of value in the treatment of venereal disease and its various symptoms, especially the skin conditions. This can only emphasize how uncritically the attributes, or lack of them, of various vegetable drugs are still assessed by their fans. There is absolutely no physiological or chemical evidence to support any claim of this kind of therapeutic ability for yellow dock. However, because of its content of tannin and anthraquinones, the drug's astringent and laxative properties are well established.

REFERENCES

1. M. Tierra. *The Way of Herbs.* Unity Press, Santa Cruz, California, 1980, pp. 121-122.

2. *The National Formulary,* Fifth Edition. American Pharmaceutical Association, Washington, DC, 1926, pp. 386-387.

3. P. H. List and L. Hörhammer (Eds.). *Hagers Handbuch der Pharmazeutischen Praxis,* Fourth Edition, Volume 6B. Springer-Verlag, Berlin, 1979, pp. 192-194.

4. J. J. Raffa Arias and C. E. Molfino. *Revista Farmaceutica* (Buenos Aires) 104: 151-155, 1962.

Yohimbe

Hast du Yohimbin im Haus,
Macht der Hausfreund dir nichts aus.
Yohimbin ist grosser Mist,
Wenn's der Hausfreund selber frisst.

At first, this bit of German doggerel might seem to have little relation to a West African tree, but it is just such a plant whose bark, known as yohimbe, has long been valued as an aphrodisiac. The tree, known as *Pausinystalia yohimba* Pierre, a member of the family Rubiaceae, is native to Cameroon, Gabon, and Congo. Its bark contains up to about 6 percent of a mixture of alkaloids, the principal one being yohimbine.[1]

Yohimbe and yohimbine enjoy a considerable folkloric reputation as aphrodisiacs, that is, drugs which stimulate sexual desire and performance. One recipe recommends boiling six to ten teaspoonfuls of inner bark shavings in a pint of water for a few minutes, straining, sweetening, and drinking the beverage. The alkaloidal salt yohimbine hydrochloride is usually administered in 5.4-mg doses. It is available as a prescription drug in a variety of combinations with other so-called sexual stimulants, including strychnine, thyroid, and methyltestosterone. Some authors recommend snuffing yohimbine to obtain both stimulant and mild hallucinogenic effects.[2]

The drug dilates the blood vessels of the skin and mucous membranes but simultaneously increases blood pressure. Its alleged aphrodisiac effects are attributed not only to this enlargement of blood vessels in the sexual organs but to increased reflex excitability in the sacral (lower) region of the spinal cord. Early scientific studies of the aphrodisiac properties of yohimbine had produced unimpressive results. Then, in 1984, an investigation of the effect of relatively small doses in sexually active male rats concluded that the drug definitely increased sexual arousal in the treated animals.[3] The investigators concluded that these results differed from those of earlier studies because the much larger doses of the drug previously employed produced other behavioral changes in the test animals. It is not yet possible to draw any firm conclusions regarding the aphrodisiac effects of yohimbe and yohimbine in human beings, but several studies give positive indications of their value.[4,5]

Yohimbe is a weak monoamine oxidase inhibitor, but it also increases monoamine production, so its overall effect is that of a reasonably active MAO inhibitor. This means that tyramine-containing foods (liver, cheese, red wine, etc.) and nasal decongestants or certain diet aids containing phenylpropanolamine should be rigorously avoided if it is used. The drug also should not be taken by persons suffering from hypotension, diabetes, or from heart, liver, or kidney disease. Psychic reactions resembling anxiety have been shown to be produced by yohimbine. In the case of individuals suffering from schizophrenia, it may actually activate psychoses.[6,7] These unpleasant and potentially hazardous reactions make it impossible to recommend the use of yohimbe for self-treatment. In any event, neither it nor yohimbine is now readily available over the counter in the United States, the latter having been declared both unsafe and ineffective by the FDA for over-the-counter sale.

German health authorities also do not recommend the therapeutic use of yohimbe for two reasons: insufficient proof of its effectiveness and a risk of serious side effects.[8] Despite this finding, yohimbe-containing preparations are available in every "sex shop" in Germany and are frequently advertised in various popular magazines there. Yohimbe products continue to be sold in the United States as dietary supplements, but they have been shown to be of extremely poor quality. Many were essentially devoid of yohimbine and lacked other alkaloids that would be expected to be present in authentic yohimbe bark.[9] In both America and Britain, yohimbine tablets are often the subject of advertisements in reputable medical journals. It is unfortunate that an herbal product which has shown some promise of value in the treatment of psychogenic impotence and which is so readily available and apparently widely used, at least in Europe, should be so soundly condemned by health authorities, both here and abroad.

Since the new prescription drug Viagra has recently supplanted virtually all other treatments for impotence with success seldom seen before in any drug, either man-made or derived from nature, there is probably little hope for any new research on yohimbe. Still, it would be nice to know more about its safety and efficacy.

For those who cannot read German and are still curious, the author's English version of the introductory poem reads:

> If in the house there's yohimbine about,
> Your wife's secret lover just won't make out.
> But be careful, it may do you no good at all,
> If he finds it and takes it, then he'll have the ball.

REFERENCES

1. E. Steinegger and R. Hänsel. *Lehrbuch der Pharmakognosie,* Third Edition. Springer-Verlag, Berlin, 1972, p. 327-328.

2. L. A. Young, L. G. Young, M. M. Klein, D. M. Klein, and D. Beyer. *Recreational Drugs.* Collier Books, New York, 1977, pp. 207-208.

3. V. E. Tyler. *Pharmacy International* 7: 203-207, 1986.

4. Anon. *Lancet* II: 1194-1195, 1986.

5. K. Reid, A. Morales, C. Harris, D. H. C. Surridge, M. Condra, J. Owen, and J. Fenemore. *Lancet* II: 421-423, 1987.

6. G. Holmberg and S. Gershon. *Psychopharmacologia* 2: 93-106, 1961.

7. C. G. Ingram. *Clinical Pharmacology and Therapeutics* 3: 345-352, 1962.

8. *Bundesanzeiger,* August 14, 1987; February 1, 1990.

9. J. Betz, K. D. White, A. H. Der Marderosian. *Journal of AOAC International* 78: 1189-1194, 1995.

Yucca

About forty species of the genus *Yucca* grow in the warmer parts of North America, and a few species are hardy in colder climates. These members of the family Agavaceae are extensively cultivated, particularly in the South. The plants, with their stiff, usually sword-shaped leaves, may or may not have an erect, central stem. Many have descriptive common names that are much more widely recognized than their botanical designations. *Yucca aloifolia* L. is called Spanish-bayonet or dagger plant; *Y. brevifolia* Engelm. is the well-known Joshua tree; *Y. glauca* Nutt. is referred to as soapweed; *Y. whipplei* Torr. is our-Lord's-candle.[1] *Yucca schidigera* Roezl. *ex* Ortiges, Mohave yucca, is also common in the herb trade.

Yucca species, together with other agaves, are known to contain large quantities of saponins. These bitter, generally irritating principles are characterized by their capacity to foam when shaken with water. The saponins in yucca are steroid derivatives and have been extensively studied because of their potential ability as starting materials for the synthesis of cortisone and related corticoids. The specific identity and the amounts of the numerous saponins in yucca were found to vary markedly with the part of the plant tested and the season when it was collected.[2]

The recommendation of yucca in medicine, unlike that of many plant materials we have considered, is of relatively recent origin. It stems from 1975 when the results of a study titled "Yucca Plant Saponin in the Management of Arthritis" appeared in *The Journal of Applied Nutrition.*[3] Essentially, the investigators claimed to have shown that a "saponin extract" of the "desert yucca plant," taken four times daily, was both safe and effective in treating the various forms of arthritis. Neither the species nor the plant part from which the saponin extract was obtained was revealed, nor was its method of preparation specified.

The Arthritis Foundation has analyzed the methodology and results of this study and pointed out a number of deficiencies. The investigators did not differentiate between rheumatoid arthritis and osteoarthritis, two very different diseases. Other medications, in addition to the yucca, continued to be taken by the patients. Individual dosages and lengths of treatment were very different (one week to fifteen months), but results were all lumped together. Most patient response was subjective and not based on physical evidence. Some of the reported results were inconsistent. Along with these objections add the unknown composition of the drug itself and the lack of assured uniformity in different lots.

Charles C. Bennett, Vice President of Public Education for the Arthritis Foundation, has suggested that inquiries concerning yucca be answered ". . . by saying that there is no proper scientific evidence that yucca tablets are helpful in treating rheumatoid arthritis or osteoarthritis; that they are probably harmless; and that the real danger would be in taking yucca tablets INSTEAD OF following proper and proven treatment procedures, which could lead to irreversible joint damage and possible disabilities."[4] Nothing need be added to this statement.

Hope may be on the horizon for a revival in economic interest in yucca. A recent study found that extracts of *Y. schidigera* greatly reduced fecal odor when added to cat and dog food (as rated by a panel of experienced olfactory observers).[5]

REFERENCES

1. L. H. Bailey and E. Z. Bailey. *Hortus Third.* Macmillan, New York, 1976, pp. 1178-1179.

2. R. Hegnauer. *Chemotaxonomie der Pflanzen,* Volume 2. Birkhäuser Verlag, Basel, Switzerland, 1963, pp. 27-36.

3. R. Bingham, B. A. Bellew, and J. G. Bellew. *Journal of Applied Nutrition* 27(2-3): 45-51, 1975.

4. C. C. Bennett. Public Information Memo. The Arthritis Foundation, New York, February 22, 1977.

5. J. A. Lowe and S. J. Kershaw. *Research in Veterinary Science* 63: 61-66, 1997.

Summarized Evaluation of Herbal Remedies

The following section consists primarily of a chart providing a summarized evaluation of the various herbal remedies and related products discussed in detail in each monograph of the previous section. The summary includes both the common and scientific names of the plant, the part used, the principal uses, its apparent effectiveness, and its probable safety.

We must add a word of caution regarding the last two categories, which, in our experience, are unique inclusions in an herbal intended for popular consumption. The value judgments presented are those formed by the authors after detailed study of literally thousands of books and papers devoted to natural products used as drugs. In most cases, they are not based on the outcomes of the extensive double-blind clinical studies in human beings that the Food and Drug Administration requires to prove a drug safe and effective. Rather, they are founded on the majority of satisfactory evidence obtained from all sources regarding each drug.

For example, in the United States, German or Hungarian chamomile has not been declared a safe and effective drug, at least as far as the FDA is concerned. The reason for this is that the FDA has not been presented with sufficient evidence to prove unequivocally its safety and efficacy as a medicinal agent. That is because no drug company is willing to spend the money to obtain this evidence since they probably could not profit from such an investment.

Still, various types of chamomile have been used as carminative, anti-inflammatory, antispasmodic, and anti-infective agents since the time of the Egyptians. Under its Latin title *Matricaria,* German chamomile was granted official status first in *The United States Pharmacopeia* and then in *The National Formulary* for an extended period (108 years). This status did not end until 1950. With the exception of an infrequent allergic reaction in sensitive individuals, reports of untoward side effects from chamomile are essentially lacking in the literature. The crude plant material is "Generally Recognized as Safe" when used as a spice, seasoning, or flavoring agent. In these capacities, it appears on the so-called GRAS list of the Food and Drug Administration.

Scores of medicinal chamomile products, ranging from the crude drug incorporated in various medicinal tea mixtures to preparations containing the purified volatile oil, are marketed in European countries. Scientific papers from those countries report favorable therapeutic results with chamomile in both small animals and human beings. Commission E of the German Federal Department of Health has reviewed all of the data and declared chamomile to be an effective drug with anti-inflammatory, antispasmodic, wound-healing, deodorizing, and antibacterial properties. Further, it reported no known contraindications or side effects. For all of these reasons, in the following table, chamomile has been rated as an apparently safe and probably effective drug, when used appropriately.

All of the value judgments given in the table, both positive and negative, are based on similar evidence and reasoning. If used with this understanding, the chart will provide the busy reader at a single glance the essential information about each herb included in this book.

The chart does not comment on the desirability or feasibility of using any of these remedies, even those indicated as being apparently efficacious and probably safe. The interested reader should consult the text for detailed explanations of the many complex factors regarding the use of these herbs, which could not be included in this brief summary.

SUMMARY CHART

Common Name	Source	Part Used
Alfalfa	*Medicago sativa*	leaves and tops
Aloe	*Aloë barbadensis*	1. fresh gel 2. dried juice
Angelica	*Angelica archangelica*	root, fruit, leaves
Apricot Pits (Laetrile)	*Prunus armeniaca*	seed kernels
Arnica	*Arnica* spp.	flower heads
Barberry	*Mahonia* and *Berberis* spp.	rhizome and roots
Bayberry	*Myrica pensylvanica* and *M. cerifera*	1. root bark 2. berries
Betony	*Stachys officinalis*	leaves and tops
Black Cohosh	*Cimicifuga racemosa*	rhizome and roots
Blue Cohosh	*Caulophyllum thalictroides*	rhizome and roots
Boneset	*Eupatorium perfoliatum*	leaves and tops
Borage	*Borago officinalis*	leaves and tops
Bran	*Triticum aestivum*	outer seed coat
Broom	*Cytisus scoparius*	flowering tops
Buchu	*Barosma* spp.	leaves
Burdock	*Arctium* spp.	root
Butcher's-Broom	*Ruscus aculeatus*	rhizome and root

Principal Uses	Apparent Efficacy[a]	Probable Safety[a]
antiarthritic, lower cholesterol	–	+
wound healing, burns cathartic	+ +	+ +
1. antiflatulent, emmenagogue, etc. 2. flavor	+ +	– +
anticancer	–	–
anti-inflammatory, analgesic	+	+
antibacterial and astringent	+	+
astringent (diarrhea) fragrant wax candles	+ +	±[b] +
astringent (diarrhea, sore throat)	+	±[b]
antirheumatic, uterine problems, etc.	±	±
uterine stimulant, emmenagogue, etc.	+	–
1. break up colds and flu 2. induce sweating	– +	+ +
diuretic, astringent (diarrhea)	– to ±	±
increase dietary fiber, benefit various gastrointestinal conditions	+	+
mind-altering properties (smoked)	+	–
urinary antiseptic, diuretic	± to +	+
alterative, treatment of skin disorders	–	+
improve circulation	+	+

Common Name	Source	Part Used
Caffeine-Containing Plants		
Coffee	*Coffea arabica*	seeds
Tea	*Camellia sinensis*	leaves and leaf buds
Kola	*Cola nitida*	cotyledons (seed leaves)
Cacao	*Theobroma cacao*	seeds
Guarana	*Paullinia cupana*	seeds
Maté	*Ilex paraguariensis*	seeds leaves
Calamus	*Acorus calamus*	rhizome
Calendula (Marigold)	*Calendula officinalis*	ligulate florets (flower parts)
Canaigre	*Rumex hymenosepalus*	root
Capsicum	*Capiscum* spp.	fruits
Catnip	*Nepeta cataria*	leaves and tops
Cat's Claw	*Uncaria tomentosa* and *U. guianensis*	root bark and stem bark
Celery Seed	*Apium graveolens*	seeds and leaf stalk
Chamomiles and Yarrow	*Matricaria recutita* *Chamaemelum nobile* *Achillea millefolium*	flower heads flower heads flowering herb
Chaparral	*Larrea tridentata*	leaves and twigs
Chickweed	*Stellaria media*	leaves and stems
Chicory	*Cichorium intybus*	root
Coltsfoot	*Tussilago farfara*	leaves and/or flower heads
Comfrey	*Symphytum officinale* and *S.* x *uplandicum*	rhizome and roots, leaves

Principal Uses	Apparent Efficacy[a]	Probable Safety[a]
	+	±
central stimulant		
febrifuge, digestive aid	±	− or ±
facilitate wound healing	±	+
tonic	−	−[b]
rubefacient, stomachic, chronic pain	+	+
1. digestive aid, sleep aid 2. mind-altering properties (smoked)	± −	+ ±
1. immunostimulant 2. folk remedy for cancer	± −	± −
anti-inflammatory	−	+
carminative, anti-inflammatory, antispasmodic, and anti-infective	+	+
alterative, anticancer	−	−
treatment of skin disorders, various internal ailments	−	+
caffeine-free beverage	+	+
antitussive (coughs), demulcent	+	−
general healing agent	+	−

Common Name	Source	Part Used
Cranberry	*Vaccinium macrocarpon*	fruit
Cucurbita	*Cucurbita* spp.	seeds
Damiana	*Turnera diffusa* var. *aphrodisiaca*	leaves
Dandelion	*Taraxacum officinale*	1. rhizome and roots 2. leaves
Devil's Claw	*Harpagophytum procumbens*	secondary storage roots
Dong Quai	*Angelica sinensis*	root
Echinacea	*Echinacea angustifolia*	rhizome and roots, overground plant
Ephedra (Ma Huang)	*Ephedra* spp. (other than Central or North American origin)	stems
Evening Primrose	*Oenothera biennis*	seed oil
Eyebright	*Euphrasia officinalis*	entire overground plant
Fennel	*Foeniculum vulgare*	fruit (seeds)
Fenugreek	*Trigonella foenumgraecum*	seeds
Feverfew	*Tanacetum parthenium*	leaves
Fo-Ti (He-Shou-Wu)	*Polygonum multiflorum*	tuberous root
Garcinia	*Garcinia cambogia*	Fresh fruit extracts
Garlic and Other Alliums	*Allium sativum* (garlic) *Allium cepa* (onion) *Allium ampeloprasum* (leek) *Allium ascalonicum* (scallion)	bulbs and occasionally leaves

Principal Uses	Apparent Efficacy[a]	Probable Safety[a]
treatment of urinary tract infections	+	+
teniafuge (expel intestinal worms), treatment of prostatic hypertrophy	+ –	+ +
aphrodisiac	–	+
digestive aid, laxative diuretic	± ±	+ +
antirheumatic	–	+
uterine tonic, antispasmodic, alterative	±	–
anti-infective, wound healing, immune stimulant	+	+
antiasthmatic, nasal decongestant	+	±
treatment of atopic eczema, mastalgia	±	±
treatment of eye diseases (conjunctivitis)	–	–
stomachic, carminative	+	+
1. demulcent, stomachic 2. flavor	± +	+ +
migraine preventive	+	+
1. cathartic 2. rejuvenation	+ –	+ +
weight loss	–	–
treatment of atherosclerosis and high blood pressure, blood clotting disorders, gastrointestinal ailments	+	+

Common Name	Source	Part Used
Gentian	*Gentiana lutea*	rhizome and roots
Ginger	*Zingiber officinale*	rhizome
Ginkgo	*Ginkgo biloba*	leaf extract
Ginseng and Related Herbs	*Panax ginseng* (Oriental ginseng) *Panax quinquefolius* (American ginseng) *Panax pseudo-ginseng* (San qui ginseng) *Eleutherococcus senticosus* (Eleuthero)	roots
Goldenseal	*Hydrastis canadensis*	rhizome and roots
Gotu Kola	*Centella asiatica*	leaves
Grape Seed Extract	*Vitis vinifera*	seed extract
Hawthorn	*Crataegus laevigata*	fruits (haws), leaves, flowers
Hibiscus	*Hibiscus sabdariffa*	flowers
Honey	*Apis mellifera*	saccharine secretion
Hops	*Humulus lupulus*	fruits (strobiles)
Horehound	*Marrubium vulgare*	leaves and tops
Horsetail	*Equisetum arvense*	overground plant
Hydrangea	*Hydrangea arborescens* *Hydrangea paniculata*	rhizome and roots leaves
Hyssop	*Hyssopus officinalis*	leaves and tops

Principal Uses	Apparent Efficacy[a]	Probable Safety[a]
appetite stimulant, digestive aid	+	+
motion sickness preventive	+	+
enhance cerebral blood flow	+	+
adaptogen, tonic, cure-all, antistress agent	±	+
bitter tonic, digestive aid, treatment of genitourinary disorders	± to +	+
promote longevity, aphrodisiac	–	+
antioxidant	+	+
dilate blood vessels, strengthen heart, lower blood pressure	+	+
laxative, diuretic	± to +	+
1. antiarthritic, sedative 2. nutrient, sweetener	– +	+ +
1. sedative, sleep aid 2. mind-altering action	± to + ±	+ ±
expectorant (coughs)	+	+
diuretic and astringent in kidney and bladder ailments	– to ±	+
1. diuretic and treatment of kidney stones 2. mind-altering action (smoked)	– +	+ –
expectorant (coughs and colds)	+	+

Common Name	Source	Part Used
Jojoba Oil	*Simmondsia chinensis*	expressed from seeds
Juniper	*Juniperus communis*	fruits (berries)
Kava	*Piper methysticum*	rootstock
Kelp	*Laminaria, Macrocystis, Nereocystis,* and *Fucus* spp.	entire plant
Lettuce Opium	*Lactuca virosa* and related species	dried latex
Licorice	*Glycyrrhiza glabra*	rhizome and roots
Life Root	*Senecio aureus*	entire plant
Linden Flowers	*Tilia* spp.	flowers
Lobelia	*Lobelia inflata*	leaves and tops
Lovage	*Levisticum officinale*	1. rhizome and roots 2. leaves
Milk Thistle	*Silybum marianum*	fruits (seeds)
Mistletoe	*Phoradendron leucarpum* (American mistletoe) *Viscum album* and subspp. (European mistletoe)	leaves leaves
Mormon Tea	*Ephedra nevadensis*	stems
Muira Puama (Potency Wood)	*Ptychopetalum olacoides* and *P. uncinatum*	stem wood, root
Mullein	*Verbascum thapsus*	leaves, flowers

Principal Uses	Apparent Efficacy[a]	Probable Safety[a]
1. antisebum shampoos 2. emollient lotions, cosmetics	± +	+ +
diuretic	+	±
antianxiety	+	+[a]
1. bulk laxative, demulcent 2. control obesity, atherosclerosis	+ –	± ±
1. sedative, analgesic 2. mind-altering action (smoked)	– –	+ ±
expectorant, demulcent, flavor	+	+ to –
emmenagogue, treatment of uterine diseases	±	–
diaphoretic, beverage	+	+
1. nauseant expectorant 2. mind-altering action	+ +	± –
diuretic, carminative flavor	+ +	+ +
liver protectant	+	+
stimulate smooth muscle, increase blood pressure antispasmodic, reduce blood pressure	± ±	– –
1. alterative, tonic 2. diuretic, astringent (diarrhea)	– +	±[b] ±[b]
aphrodisiac	–	±
demulcent, emollient, astringent	+	+

Common Name	Source	Part Used
Myrrh	*Commiphora* spp.	oleo-gum-resin
Nettle	*Urtica dioica*	overground plant
New Zealand Green-Lipped Mussel	*Perna canaliculus*	entire organism
Pangamic Acid (Vitamin B_{15})	*Prunus armeniaca* or synthetic	chemical or chemical mixture
Papaya	*Carica papaya*	1. dried latex 2. leaves
Parsley	*Petroselinum crispum*	1. leaves and stems 2. fruit (seeds)
Passion Flower	*Passiflora incarnata*	flowering and fruiting top
Pau d'Arco	*Tabebuia* spp.	bark
Pennyroyal	*Hedeoma pulegioides* (American pennyroyal) *Mentha pulegium* (European pennyroyal)	leaves oil
Peppermint	*Mentha* x *piperita*	leaves
Pokeroot	*Phytolacca americana*	root
Pollen	seed-bearing plants	microspores (male reproductive elements)
Propolis	bee hives—originally from conifer and poplar trees	resinous material
Pygeum	*Prunus africana*	bark

Principal Uses	Apparent Efficacy[a]	Probable Safety[a]
astringent, protective, fragrance	+	+
1. diuretic 2. treatment of prostatic hypertrophy 3. antiasthmatic, antirheumatic, stimulate hair growth	+ ± −	+ + +
antiarthritic	−	+
various, detoxify poisonous products in human system	−	− to ∓
digestive aid, vermifuge	−	+
digestive aid, diuretic digestive aid, diuretic, emmenagogue	± +	+ ± to +
sedative, calmative	± to +	+
anticancer	−	±
carminative, diaphoretic, emmenagogue emmenagogue, abortifacient	± ± to +	+ −
stomachic, carminative, flavor	+	+
alterative, antirheumatic, anticancer, cathartic, etc.	−	−
tonic, treatment of various debilitating conditions	−	− to ∓
antibacterial activity (tuberculosis), gastrointestinal disturbances	±	+
treatment of prostatic hypertrophy	+	+

Common Name	Source	Part Used
Raspberry	*Rubus idaeus* and *R. strigosus*	leaves
Red Bush Tea	*Aspalathus linearis*	leaves and fine twigs
Red Clover	*Trifolium pratense*	flowers
Rose Hips	*Rosa* spp.	fruits
Rosemary	*Rosmarinus officinalis*	leaves and/or tops
Royal Jelly	*Apis mellifera*	pharyngeal gland secretion
Rue	*Ruta graveolens*	leaves
Sage	*Salvia officinalis*	leaves
St. John's Wort	*Hypericum perforatum*	leaves and tops
Sarsaparilla	*Smilax* spp.	roots
Sassafras	*Sassafras albidum*	root bark
Savory	*Satureja hortensis* (summer savory)	overground plant
	Satureja montana (winter savory)	overground plant
Saw Palmetto	*Serenoa repens*	ripe fruits
Schisandra	*Schisandra chinensis*	fruits

Principal Uses	**Apparent Efficacy**[a]	**Probable Safety**[a]
astringent, stimulant, treatment of conditions associated with pregnancy	+	+[b]
refreshing beverage	+	+
alterative, anticancer treatment	–	+
antiscorbutic	+	+
tonic, diaphoretic, stomachic, antirheumatic, spice, flavor	+	+
tonic, prevent aging	–	+
antispasmodic, emmenagogue	–	–
1. astringent, anhidrotic (reduce secretions) 2. flavor	+ +	– to ± +
1. alterative 2. diuretic, flavor	 +	+ +
1. stimulant, antispasmodic, sudorific, antirheumatic, tonic 2. flavor	– +	– –
1. carminative, appetite stimulant, antidiarrhea 2. aphrodisiac, decrease sex drive	+ – –	+ + +
1. diuretic, treatment of prostatic hypertrophy 2. increase size of mammary glands, stimulate sexual vigor	+ –	+ +
stimulant, liver protectant	–	±

Common Name	Source	Part Used
Scullcap	*Scutellaria lateriflora*	overground plant
Senega Snakeroot	*Polygala senega*	root
Senna	*Cassia senna* (Alexandria senna) *Cassia angustifolia* (Tinnevelly senna)	leaflets
Spirulina	*Spirulina* spp. (blue-green algae)	entire plants
Suma	*Hebanthe paniculata*	root
Tansy	*Tanacetum vulgare*	leaves and tops
Tea Tree	*Melaleuca alternifolia*	volatile oil
L-Tryptophan	casein or bacterial metabolite	amino acid
Uva Ursi	*Arctostaphylos uva-ursi*	leaves
Valerian	*Valeriana officinalis* and related *V.* spp.	rhizome and roots
Wild Yam	*Dioscorea villosa*	rhizome and roots
Witch Hazel	*Hamamelis virginiana*	leaves, bark
Wormwood	*Artemisia absinthium*	leaves and tops
Yellow Dock	*Rumex crispus*	rhizome and roots
Yohimbe	*Pausinystalia yohimba*	bark
Yucca	*Yucca* spp.	leaves

+ = Effective, safe in normal individuals.
± = Efficacy or safety inconclusive.
– = Ineffective, not safe.
a = When used appropriately. See specific monograph for details or administration.
b = All tannin-rich drugs may have carcinogenic potential in long-term usage.
c = Due to impurities.

Principal Uses	Apparent Efficacy[a]	Probable Safety[a]
tonic, tranquilizing effects, antispasmodic	–	±
expectorant, diaphoretic, emetic	+	+
cathartic	+	+
1. nutrient 2. appetite suppressant	+ –	+ +
adaptogen, cure-all	–	±
anthelmintic, tonic, emmenagogue	±	– to ±
antiseptic	+	+
sleep aid, antidepressant	+	–[c]
diuretic, urinary antiseptic, astringent	+	+
tranquilizer, calmative	+	+
"natural alternative" to estrogen replacement therapy	–	–
astringent	+	+
anthelmintic, tonic, mind-altering action, flavor	+	–
astringent, laxative	+	+[b]
aphrodisiac, sexual stimulant	+	– to ±
antiarthritic	–	+

Index